Intelligent Systems Reference Library

Volume 114

About this Series

The aim of this series is to publish a Reference Library, including novel advances and developments in all aspects of Intelligent Systems in an easily accessible and well structured form. The series includes reference works, handbooks, compendia, textbooks, well-structured monographs, dictionaries, and encyclopedias. It contains well integrated knowledge and current information in the field of Intelligent Systems. The series covers the theory, applications, and design methods of Intelligent Systems. Virtually all disciplines such as engineering, computer science, avionics, business, e-commerce, environment, healthcare, physics and life science are included.

More information about this series at http://www.springer.com/series/8578

Anand Jayant Kulkarni · Ganesh Krishnasamy
Ajith Abraham

Cohort Intelligence: A Socio-inspired Optimization Method

Springer

Anand Jayant Kulkarni
Odette School of Business
University of Windsor
Windsor, ON
Canada

and

Department of Mechanical Engineering, Symbiosis Institute of Technology
Symbiosis International University
Pune, Maharashtra
India

Ganesh Krishnasamy
Department of Electrical Engineering, Faculty of Engineering
Universiti Malaya
Kuala Lumpur
Malaysia

Ajith Abraham
Machine Intelligence Research Labs (MIR Labs)
Scientific Network for Innovation and Research Excellence
Auburn, WA
USA

ISSN 1868-4394 ISSN 1868-4408 (electronic)
Intelligent Systems Reference Library
ISBN 978-3-319-83022-3 ISBN 978-3-319-44254-9 (eBook)
DOI 10.1007/978-3-319-44254-9

Printed on acid-free paper

This Springer imprint is published by Springer Nature
The registered company is Springer International Publishing AG
The registered company address is: Gewerbestrasse 11, 6330 Cham, Switzerland

Anand Jayant Kulkarni *would like to dedicate this book to his*
loving wife 'Prajakta'
and
lovely son 'Nityay'

Preface

This book is written for engineers, scientists, and students studying/working in the optimization, artificial intelligence (AI), or computational intelligence arena. The book discusses the core and underlying principles and analysis of the different concepts associated with an emerging socio-inspired AI optimization tool referred to as cohort intelligence (CI).

The book in detail discusses the CI methodology as well as several modifications for solving a variety of problems. The validation of the methodology is also provided by solving several unconstrained test problems. In order to make CI solve real-world problems which are inherently constrained, CI method with a penalty function approach is tested on several constrained test problems and comparison of the performance is also discussed. The book also demonstrates the ability of CI methodology for solving several cases of the combinatorial problems such as traveling salesman problem (TSP) and knapsack problem (KP). In addition, real-world applications of the CI methodology by solving complex and large-sized combinatorial problems from the healthcare, inventory, supply chain optimization, and cross-border transportation domain is also discussed. The inherent ability of handling constraints based on the probability distribution is also revealed and proved using these problems. A detailed mathematical formulation, solutions, and comparisons are provided in every chapter. Moreover, the detailed discussion on the CI methodology modifications for solving several problems from the machine learning domain is also provided.

The mathematical level in all the chapters is well within the grasp of the scientists as well as the undergraduate and graduate students from the engineering and computer science streams. The reader is encouraged to have basic knowledge of probability and mathematical analysis. In presenting the CI and associated modifications and contributions, the emphasis is placed on the development of the fundamental results from basic concepts. Numerous examples/problems are worked out in the text to illustrate the discussion. These illustrative examples may allow the reader to gain further insight into the associated concepts. The various algorithms for solving have been coded in MATLAB software. All the executable codes are available online at www.sites.google.com/site/oatresearch/cohort-intelligence.

The book is an outgrowth of the three-year work by the authors. In addition, Fazle Baki and Ben Chaouch from University of Windsor, ON, Canada, helped with the complex combinatorial problem formulations. Over the period of 3 years, the algorithms have been tested extensively for solving various real-world problems as well as published in various prestigious journals and conferences. The suggestions and criticism of various reviewers and colleagues had a significant influence on the way the work has been presented in this book. We are much grateful to our colleagues for reviewing the different parts of the manuscript and for providing us valuable feedback. The authors would like to thank Dr. Thomas Ditzinger, Springer Engineering In-house Editor, Studies in Computational Intelligence Series; Prof. Janusz Kacprzyk, Editor-in-Chief, Springer Intelligence Systems Reference Library Series; and Mr. Holger Schäpe, Editorial Assistant, Springer Verlag, Heidelberg, for the editorial assistance and excellent cooperative collaboration to produce this important scientific work. We hope that the reader will share our excitement to present this volume on *cohort intelligence* and will find it useful.

Windsor, ON, Canada
Kuala Lumpur, Malaysia
Auburn, WA, USA
May 2016

Anand Jayant Kulkarni
Ganesh Krishnasamy
Ajith Abraham

Contents

Chapter 1
Introduction to Optimization

1.1 What Is Optimization?

For almost all the human activities there is a desire to deliver the most with the least. For example in the business point of view maximum profit is desired from least investment; maximum number of crop yield is desired with minimum investment on fertilizers; maximizing the strength, longevity, efficiency, utilization with minimum initial investment and operational cost of various household as well as industrial equipments and machineries. To set a record in a race, for example, the aim is to do the fastest (shortest time).

The concept of optimization has great significance in both human affairs and the laws of nature which is the inherent characteristic to achieve the best or most favorable (minimum or maximum) from a given situation [1]. In addition, as the element of design is present in all fields of human activity, all aspects of optimization can be viewed and studied as design optimization without any loss of generality. This makes it clear that the study of design optimization can help not only in the human activity of creating optimum design of products, processes and systems, but also in the understanding and analysis of mathematical/physical phenomenon and in the solution of mathematical problems. The constraints are inherent part if the real world problems and they have to be satisfied to ensure the acceptability of the solution. There are always numerous requirements and constraints imposed on the designs of components, products, processes or systems in real-life engineering practice, just as in all other fields of design activity. Therefore, creating a feasible design under all these diverse requirements/constraints is already a difficult task, and to ensure that the feasible design created is also 'the best' is even more difficult.

A.J. Kulkarni et al., *Cohort Intelligence: A Socio-inspired Optimization Method*,
Intelligent Systems Reference Library 114, DOI 10.1007/978-3-319-44254-9_1

1.1.1 General Problem Statement

All the optimal design problems can be expressed in a standard general form stated as follows:

$$\text{Minimize objective function} \quad f(\mathbf{X}) \tag{1.1}$$

Subject to

$$s \text{ number of inequality constraints } g_j(\mathbf{X}) \leq 0, \quad j = 1, 2, \ldots, s \tag{1.2}$$

$$w \text{ number of equality constraints } \quad h_j(\mathbf{X}) = 0, \quad j = 1, 2, \ldots, w \tag{1.3}$$

where the number of design variables is given by $x_i, \quad i = 1, 2, \ldots, n$

or by design variable vector $\mathbf{X} = \begin{Bmatrix} x_1 \\ x_2 \\ \vdots \\ x_n \end{Bmatrix}$

- A problem where the objective function is to be maximized (instead of minimized) can also be handled with this standard problem statement since maximization of a function $f(\mathbf{X})$ is the same as minimizing the negative of $f(\mathbf{X})$.
- Similarly, the '≥' type of inequality constraints can be treated by reversing the sign of the constraint function to form the '≤' type of inequality.
- Sometimes there may be simple limits on the allowable range of value a design variable can take, and these are known as side constraints:

$$x_i^l \leq x_i \leq x_i^u$$

- where x_i^l and x_i^u are the lower and upper limits of x_i, respectively. However, these side constraints can be easily converted into the normal inequality constraints (by splitting them into 2 inequality constraints).
- Although all optimal design problems can be expressed in the above standard form, some categories of problems may be expressed in alternative specialized forms for greater convenience and efficiency.

1.1.2 Active/Inactive/Violated Constraints

The constraints in an optimal design problem restrict the entire design space into smaller subset known as the feasible region, i.e. not every point in the design space is feasible. See Fig. 1.1.

- An inequality constraint $g_j(\mathbf{X})$ is said to be violated at the point x if it is not satisfied there $\left(g_j(\mathbf{X}) \geq 0\right)$.
- If $g_j(\mathbf{X})$ is strictly satisfied $\left(g_j(\mathbf{X}) < 0\right)$ then it is said to be inactive at x.
- If $g_j(\mathbf{X})$ is satisfied at equality $\left(g_j(\mathbf{X}) = 0\right)$ then it is said to be active at x.
- The set of points at which an inequality constraint is active forms a constraint boundary which separates the feasibility region of points from the infeasible region.
- Based on the above definitions, equality constraints can only be either violated $\left(h_j(\mathbf{X}) \neq 0\right)$ or active $\left(h_j(\mathbf{X}) = 0\right)$ at any point x.
- The set of points where an equality constraint is active forms a sort of boundary both sides of which are infeasible.

1.1.3 Global and Local Minimum Points

Let the set of design variables that give rise to a minimum of the objective function $f(\mathbf{X})$ be denoted by $\mathbf{X}^*$ (the asterisk $*$ is used to indicate quantities and terms referring to an optimum point). An objective $G(\mathbf{X})$ is at its global (or absolute) minimum at the point $\mathbf{X}^*$ if:

$$f(\mathbf{X}^*) \leq f(\mathbf{X}) \quad \text{for all } \mathbf{X} \text{ in the feasible region}$$

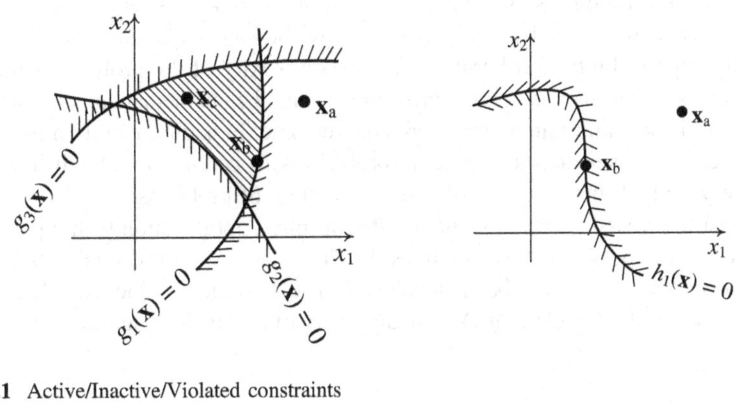

Fig. 1.1 Active/Inactive/Violated constraints

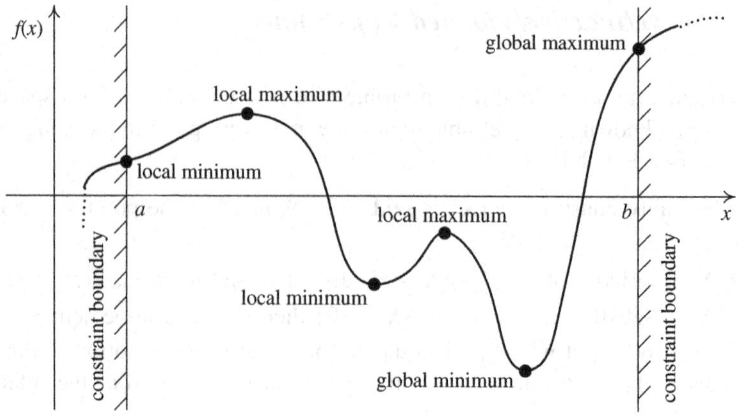

Fig. 1.2 Minimum and maximum points

The objective has a local (or relative) minimum at the point $\mathbf{X}^*$ if:

$$f(\mathbf{X}^*) \leq f(\mathbf{X}) \quad \text{for all feasible } \mathbf{X} \text{ within a small neighborhood of } \mathbf{X}^*$$

A graphical representation of these concepts is shown in Fig. 1.2 for the case of a single variable x over a closed feasible region $a \leq x \leq b$.

1.2 Contemporary Optimization Approaches

There are several mathematical optimization techniques being practiced so far, for example gradient methods, Integer Programming, Branch and Bound, Simplex algorithm, dynamic programming, etc. These techniques can efficiently solve the problems with limited size. Also, they could be more applicable to solve linear problems. In addition, as the number of variables and constraints increase, the computational time to solve the problem, may increase exponentially. This may limit their applicability. Furthermore, as the complexity of the problem domain is increasing solving such complex problems using the mathematical optimization techniques is becoming more and more cumbersome. In addition, certain heuristics have been developed to solve specific problem with certain size. Such heuristics have very limited flexibility to solve different class of problems.

In past few years a number of nature-/bio-inspired optimization techniques (also referred to as metaheuristics) such as Evolutionary Algorithms (EAs), Swarm Intelligence (SI), etc. have been developed. The EA such as Genetic Algorithm (GA) works on the principle of Darwinian theory of survival of the fittest individual in the population. The population is evolved using the operators such as selection, crossover, mutation, etc. According to Deb [2] and Ray et al. [3], GA can often reach very close to the global optimal solution and necessitates local improvement techniques to incorporate into it. Similar to GA, mutation driven approach of Differential Evolution (DE) was proposed by Storn and Price [4] which helps explore and further locally exploit the solution space to reach the global optimum. Although, easy to implement, there are several problem dependent parameters required to be tuned and may also require several associated trials to be performed.

Inspired from social behavior of living organisms such as insects, fishes, etc. which can communicate with one another either directly or indirectly the paradigm of SI is a decentralized self organizing optimization approach. These algorithms work on the cooperating behavior of the organisms rather than competition amongst them. In SI, every individual evolves itself by sharing the information from others in the society. The techniques such as Particle Swarm Optimization (PSO) is inspired from the social behavior of bird flocking and school of fish searching for food [4]. The fishes or birds are considered as particles in the solution space searching for the local as well as global optimum points. The directions of movements of these particles are decided by the best particle in the neighborhood and the best particle in entire swarm. The Ant Colony Optimization (ACO) works on the ants' social behavior of foraging food following a shortest path [5]. The ant is considered as an agent of the colony. It searches for the better solution in its close neighborhood and iteratively updates its solution. The ants also updates their pheromone trails at the end of every iteration. This helps every ant decide their directions which may further self organize them to reach to the global optimum. Similar to ACO, the Bee Algorithm (BA) also works on the social behavior of honey bees finding the food; however, the bee colony tends to optimize the use of number of members involved in particular pre-decided tasks [6]. The Bees Algorithm is a population-based search algorithm proposed by Pham et al. [7] in a technical report presented at the Cardiff University, UK. It basically mimics the food foraging behavior of honey bees. According to Pham and Castellani [8] and Pham et al. [7], Bees Algorithm mimics the foraging strategy of honey bees which look for the best solution. Each candidate solution is thought of as a flower or a food source, and a population or colony of n bees is used to search the problem solution space. Each time an artificial bee visits a solution, it evaluates its objective solution. Even though it has been proven to be effective solving continuous as well as combinatorial problems Pham and Castellani [8, 9], some measure of the topological distance between the solutions is required. The Firefly Algorithm (FA) is an emerging metaheuristic swarm optimization technique based on the natural behavior of fireflies. The natural behavior of fireflies is based on bioluminescence phenomenon [10, 11]. They produce short and rhythmic flashes to communicate with other fireflies and attract potential prey. The light intensity/brightness I of the flash at a distance r obeys inverse square law, i.e. $I \propto 1/r^2$ in addition to the light absorption by surrounding air. This makes most of the fireflies visible only till a limited distance, usually several hundred meters at night, which is enough to communicate. The flashing light of fireflies can be formulated in such a way that it is associated with the objective function to be optimized, which makes it possible to formulate optimization algorithms [10, 11]. Similar to the other metaheuristic algorithms constraint handling is one of crucial issues being addressed by researchers [12].

1.3 Socio-Inspired Optimization Domain

Every society is a collection of self interested individuals. Every individual has a desire to improve itself. The improvement is possible through learning from one another. Furthermore, the learning is achieved through interaction as well as competition with the individuals. It is important to mention here that this learning may lead to quick improvement in the individual's behavior; however, it is also possible that for certain individuals the learning and further improvement is slower. This is because the learning and associated improvement depend upon the quality of the individual being followed. In the context of optimization (minimization and maximization) if the individual solution being followed is better, the chances of improving the follower individual solution increases. Due to uncertainty, this is also possible that the individual solution being followed may be of inferior quality as compared to the follower candidate. This may make the follower individual solution to reach a local optimum; however, due to inherent ability of societal individuals to keep improving itself other individuals are also selected for learning. This may make the individuals further jump out of the possible local optimum and reach the global optimum solution. This common goal of improvement in the behavior/solution reveals the self organizing behavior of the entire society. This is an effective self organizing system which may help in solving a variety of complex optimization problems.

The following chapters discuss an emerging Artificial Intelligence (AI) optimization technique referred to as Cohort Intelligence (CI). The framework of CI along with its validation by solving several unconstrained test problems is discussed in detail. In addition, numerous applications of CI methodology and its modified versions in the domain of machine learning are provided. Moreover, the CI application for solving several test cases of the combinatorial problems such as Traveling Salesman Problem (TSP) and 0–1 Knapsack Problem are discussed. Importantly, CI methodology solving real world combinatorial problems from the healthcare and inventory problem domain, as well as complex and large sized Cross-Border transportation problems is also discussed. These applications underscore the importance of the Socio-inspired optimization method such as CI.

References

1. Kulkarni, A.J., Tai, K., Abraham, A.: Probability collectives: a distributed multi-agent system approach for optimization. In: Intelligent Systems Reference Library, vol. 86. Springer, Berlin (2015) (doi:10.1007/978-3-319-16000-9, ISBN: 978-3-319-15999-7)
2. Deb, K.: An efficient constraint handling method for genetic algorithms. Comput. Methods Appl. Mech. Eng. **186**, 311–338 (2000)
3. Ray, T., Tai, K., Seow, K.C.: Multiobjective design optimization by an evolutionary algorithm. Eng. Optim. **33**(4), 399–424 (2001)
4. Storn, R., Price, K.: Differential evolution—a simple and efficient heuristic for global optimization over continuous spaces. J. Global Optim. **11**, 341–359 (1997)
5. Kennedy, J., Eberhart, R.: Particle swarm optimization. In: Proceedings of IEEE International Conference on Neural Networks, pp. 1942–1948 (1995)
6. Dorigo, M., Birattari, M., Stitzle, T.: Ant colony optimization: artificial ants as a computational intelligence technique. IEEE Comput. Intell. Mag., 28–39 (2006)
7. Pham, D.T., Ghanbarzadeh, A., Koc, E., Otri, S., Rahim, S., Zaidi, M.: The bees algorithm. Technical Note, Manufacturing Engineering Centre, Cardiff University, UK (2005)
8. Pham, D.T., Castellani, M.: The bees algorithm—modelling foraging behaviour to solve continuous optimisation problems. Proc. ImechE, Part C, **223**(12), 2919–2938 (2009)
9. Pham, D.T., Castellani, M.: Benchmarking and comparison of nature-inspired population-based continuous optimisation algorithms. Soft Comput. 1–33 (2013)
10. Yang, X.S.: Firefly algorithms for multimodal optimization. In: Stochastic Algorithms: Foundations and Applications. Lecture Notes in Computer Sciences 5792, pp. 169–178. Springer, Berlin (2009)
11. Yang, X.S., Hosseini, S.S.S., Gandomi, A.H.: Firefly Algorithm for solving non-convex economic dispatch problems with valve loading effect. Appl. Soft Comput. **12**(3), 1180–1186 (2002)
12. Deshpande, A.M., Phatnani, G.M., Kulkarni, A.J.: Constraint handling in firefly algorithm. In: Proceedings of IEEE International Conference on Cybernetics, pp. 186–190 (2013)

Chapter 2
Socio-Inspired Optimization Using Cohort Intelligence

The nature-/bio-inspired optimization techniques such as genetic algorithm (GA), particle swarm optimization (PSO), ant colony optimization (ACO), simulated annealing (SA), Tabu search, etc., have become popular due to their simplicity to implement and working based on rules. The GA is population based which is evolved using the operators such as selection, crossover, mutation, etc. According to Deb [1] and Ray et al. [2] the performance of GA is governed by the quality of the population being evaluated and may often reach very close to the global optimal solution and necessitates local improvement techniques to incorporate into it. The paradigm of Swarm Intelligence (SI) is a decentralized self organizing optimization approach inspired from social behavior of living organisms such as insects, fishes, etc. which can communicate with one another either directly or indirectly. The techniques such as Particle Swarm Optimization (PSO) is inspired from the social behavior of bird flocking and school of fish searching for food [3]. The Ant Colony Optimization (ACO) works on the ants' social behavior of foraging food following a shortest path [4]. Similar to ACO, the Bee Algorithm (BA) also works on the social behavior of honey bees finding the food; however, the bee colony tends to optimize the use of number of members involved in particular pre-decided tasks [5]. Generally, the swarm techniques are computationally intensive.

Kulkarni et al. [6] proposed an emerging Artificial Intelligence (AI) technique referred to as Cohort Intelligence (CI). It is inspired from the self-supervised learning behavior of the candidates in a cohort. The cohort here refers to a group of candidates interacting and competing with one another to achieve some individual goal which is inherently common to all the candidates. When working in a cohort, every candidate tries to improve its own behavior by observing the behavior of every other candidate in that cohort. Every candidate may follow a certain behavior in the cohort which according to itself may result into improvement in its own behavior. As certain qualities make a particular behavior which, when a candidate follows, it actually tries to adapt to the associated qualities. This makes every candidate learn from one another and helps the overall cohort behavior to evolve. The cohort behavior could be considered saturated, if for considerable number of

A.J. Kulkarni et al., *Cohort Intelligence: A Socio-inspired Optimization Method*,
Intelligent Systems Reference Library 114, DOI 10.1007/978-3-319-44254-9_2

learning attempts the individual behavior of all the candidates does not improve considerably and candidates' behaviors become hard to distinguish. The cohort could be assumed to become successful when for a considerable number of times the cohort behavior saturates to the same behavior.

This chapter discusses the CI methodology framework in detail and further validates its ability by solving a variety of unconstrained test problems. This demonstrates its strong potential of being applicable for solving unimodal as well as multimodal problems.

2.1 Framework of Cohort Intelligence

Consider a general unconstrained problem (in the minimization sense) as follows:

$$\begin{array}{ll} \text{Minimize} & f(\mathbf{x}) = f(x_1, \ldots x_i, \ldots, x_n) \\ \text{Subject to} & \Psi_i^{lower} \le x_i \le \Psi_i^{upper} \quad , \quad i = 1, \ldots, N \end{array} \tag{2.1}$$

As a general case, assume the objective function $f(\mathbf{x})$ as the behavior of an individual candidate in the cohort which it naturally tries to enrich by modifying the associated set of characteristics/attributes/qualities $\mathbf{x} = (x_1, \ldots x_i, \ldots, x_N)$.

Having considered a cohort with number of candidates C, every individual candidate $c\,(c = 1, \ldots, C)$ belongs a set of characteristics/attributes/qualities $\mathbf{x}^c = (x_1^c, \ldots x_i^c, \ldots, x_N^c)$ which makes the overall quality of its behavior $f(\mathbf{x}^c)$. The individual behavior of each candidate c is generally being observed by itself and every other candidate (c) in the cohort. This naturally urges every candidate c to follow the behavior better than its current behavior. More specifically, candidate c may follow $f^*\left(\mathbf{x}^{(c)}\right)$ if it is better than $f^*(\mathbf{x}^c)$, i.e. $f^*\left(\mathbf{x}^{(c)}\right) < f^*(\mathbf{x}^c)$. Importantly, following a behavior $f(\mathbf{x})$ refers to following associated qualities $\mathbf{x} = (x_1, \ldots x_i, \ldots, x_N)$ with certain variations t associated with them. However, following better behavior and associated qualities is highly uncertain. This is because; there is certain probability involved by which it selects certain behavior to follow. In addition, a stage may come where the cohort behavior could become saturated. In other words, at a certain stage, there could be no improvement in the behavior of an individual candidate for a considerable number of learning attempts. Such situation is referred to as saturation stage. This makes every candidate to expand its search around the qualities associated with the current behavior being followed. The mathematical formulation of the CI methodology is explained below in detail with the algorithm flowchart in Fig. 2.1 [6, 7].

The procedure begins with the initialization of number of candidates C, sampling interval Ψ_i for each quality $x_i, i = 1, \ldots, N$, learning attempt counter $n = 1$, and the setup of sampling interval reduction factor $r \in [0, 1]$, convergence parameter $\varepsilon = 0.0001$, number of variations t. The values of C, t and v are chosen based on preliminary trials of the algorithm.

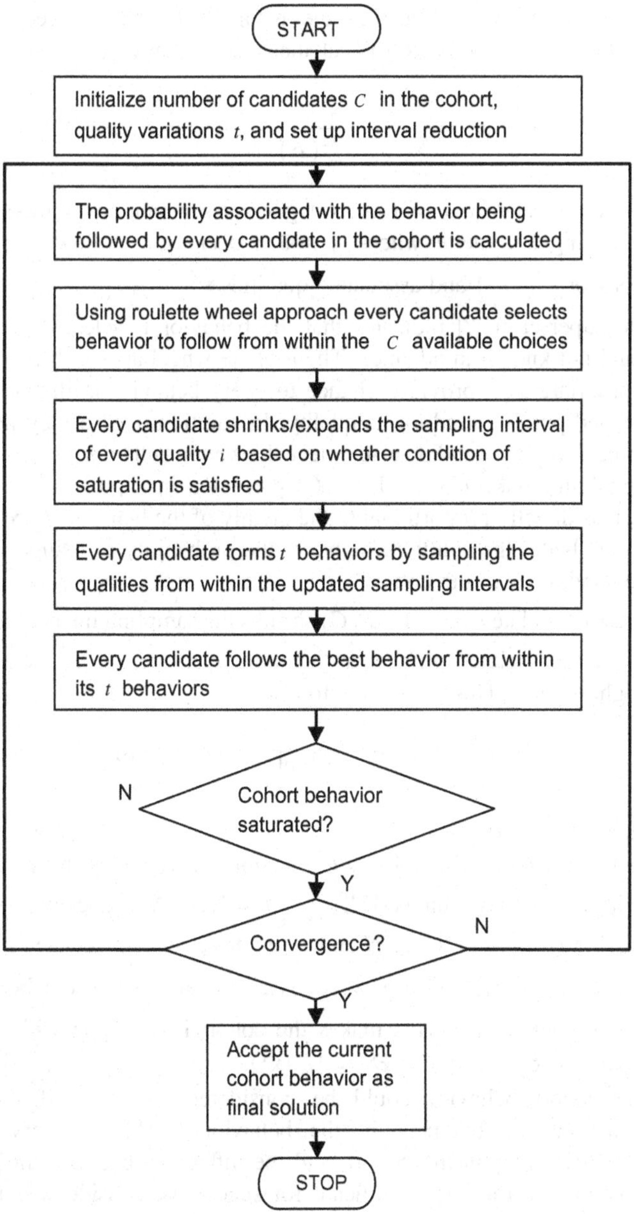

Fig. 2.1 Cohort intelligence (CI) flowchart

Step 1. The probability of selecting the behavior $f^*(\mathbf{x}^c)$ of every associated candidate $c\,(c = 1,\ldots,C)$ is calculated as follows:

$$p^c = \frac{1/f^*(\mathbf{x}^c)}{\sum_{c=1}^{C} 1/f^*(\mathbf{x}^c)}, \quad (c = 1,\ldots,C) \tag{2.2}$$

Step 2. Every candidate $c\,(c = 1,\ldots,C)$ generates a random number $r \in [0,1]$ and using a roulette wheel approach decides to follow corresponding behavior $f^*\left(\mathbf{x}^{c[?]}\right)$ and associated qualities $\mathbf{x}^{c[?]} = \left(x_1^{c[?]},\ldots x_i^{c[?]},\ldots,x_N^{c[?]}\right)$. The superscript $[?]$ indicates that the behavior is selected by candidate c and not known in advance. The roulette wheel approach could be most appropriate as it provides chance to every behavior in the cohort to get selected purely based on its quality. In addition, it also may increase the chances of any candidate to select the better behavior as the associated probability stake $p^c\,(c = 1,\ldots,C)$ presented in Eq. (2.2) in the interval $[0,1]$ is directly proportional to the quality of the behavior $f^*(\mathbf{x}^c)$. In other words, better the solution, higher is the probability of being followed by the candidates in the cohort.

Step 3. Every candidate $c\,(c = 1,\ldots,C)$ shrinks the sampling interval $\Psi_i^{c[?]},\quad i = 1,\ldots,N$ associated with every variable $x_i^{c[?]},\quad i = 1,\ldots,N$ to its local neighborhood. This is done as follows:

$$\Psi_i^{c[?]} \in \left[x_i^{c[?]} - (\|\Psi_i\|/2),\, x_i^{c[?]} + (\|\Psi_i\|/2)\right] \tag{2.3}$$

where $\Psi_i = (\|\Psi_i\|) \times r$.

Step 4. Each candidate $c\,(c = 1,\ldots,C)$ samples t qualities from within the updated sampling interval $\Psi_i^{c[?]},\quad i = 1,\ldots,N$ associated with every variable $x_i^{c[?]},\quad i = 1,\ldots,N$ and computes a set of associated t behaviors, i.e. $\mathbf{F}^{c,t} = \left\{f(\mathbf{x}^c)^1,\ldots,f(\mathbf{x}^c)^j,\ldots,f(\mathbf{x}^c)^t\right\}$, and selects the best function $f^*(\mathbf{x}^c)$ from within. This makes the cohort is available with C updated behaviors represented as $\mathbf{F}^C = \{f^*(\mathbf{x}^1),\ldots,f^*(\mathbf{x}^c),\ldots,f^*(\mathbf{x}^C)\}$.

Step 5. The cohort behavior could be considered saturated, if there is no significant improvement in the behavior $f^*(\mathbf{x}^c)$ of every candidate $c\,(c = 1,\ldots,C)$ in the cohort, and the difference between the individual behaviors is not very significant for successive considerable number of learning attempts, i.e. if

1. $\left\|\max\left(\mathbf{F}^C\right)^n - \max\left(\mathbf{F}^C\right)^{n-1}\right\| \le \varepsilon$, and
2. $\left\|\min\left(\mathbf{F}^C\right)^n - \min\left(\mathbf{F}^C\right)^{n-1}\right\| \le \varepsilon$, and

3. $\|\max(\mathbf{F}^C)^n - \min(\mathbf{F}^C)^n\| \le \varepsilon$, every candidate $c\,(c = 1, \ldots, C)$ expands the sampling interval $\Psi_i^{c[?]}$, $i = 1, \ldots, N$ associated with every quality $x_i^{c[?]}$, $i = 1, \ldots, N$ to its original one $\Psi_i^{lower} \le x_i \le \Psi_i^{upper}$, $i = 1, \ldots, N$.

Step 6. If either of the two criteria listed below is valid, accept any of the C behaviors from current set of behaviors in the cohort as the final objective function value $f^*(\mathbf{x})$ as the final solution and stop, else continue to Step 1.

(a) If maximum number of attempts exceeded.
(b) If cohort saturates to the same behavior (satisfying the conditions in Step 5) for $\tau_{\max}$ times.

2.2 Theoretical Comparison with Contemporary Techniques

Particle swarm optimization PSO is a population-based stochastic search algorithm developed by Kennedy and Eberhart [3]. Due to its simple concept, it has been applied to many optimization problems. The PSO itself did not work well in solving constrained problems [8]. To overcome this shortcoming, many modified PSO techniques such as Quantum-behaved PSO [9], Improved Vector PSO (IVPSO) [10] and other techniques which controlled the velocity of the swarm were used. All these techniques depend upon the control parameters in the velocity updating model, which are the inertia weight and acceleration coefficients. Another technique referred to as the Barebones PSO (BPSO) [11] used Gaussian normal distribution to update the values of the particles in the solution. This removed the necessity of inertia weight and acceleration coefficients. In the PSO variations the entire swarm has a collective intelligence. While the individual particle keeps track of its own best solution, every particle in the swarm is also aware of the best solution found by the entire swarm [3]. The movement of each particle is some function of this individual best and the group's best values. The Fully Informed PSO (FIPSO) is one technique that does not merely depend on the best solution offered globally. This technique samples the solutions offered by its entire neighborhood and follows a point in space that is calculated using this complete information [12].

Another technique popular today is the Genetic Algorithm (GA). This technique follows the principle of survival of the fittest. The best solutions in a particular generation are taken forward to the next one, and new solutions are generated by using crossover as well as applying mutations on it. Ant Colony optimization (ACO) [4, 13] follows autocatalytic behavior which is characterized by a positive feedback, where the probability with which an agent chooses a path increases with the number of agents that previously chose the same path [14]. However it is difficult to solve continuous optimization problem using ACO directly, as there are limitations in number of choices for ants at each stage. Some recent research has extended classical ACO to solve continuous optimization problem.

In CI, however, the collective effort of the swarm is replaced by the competitive nature of a cohort. Every candidate tries to follow the behavior of a candidate that has shown better results in that particular iteration. In following this behavior, it tries to incorporate some of the qualities that made that behavior successful. This competitive behavior motivates each candidate to perform better, and leads to an eventual improvement in the behaviors of all the candidates. This technique differs from PSO and barebones PSO in that it does not check merely the best solution, but is fully informed of the activities of its fellow candidates and follows a candidate selected using the roulette wheel approach. However, it is also different from the FIPSO which keeps track of the entire swarm in that it follows the behavior of only one candidate, not a resultant of the results presented by the entire swarm. CI also differs from GA, as there is no direct exchange of certain properties or even mutation. Rather, candidates decide to follow a fellow candidate, and try to imbibe the qualities that led that candidate to reach its solution. The values of these qualities are not replicated exactly. They are instead taken from a close neighborhood of the values of the qualities of the candidate being followed. This gives a variation in the solutions obtained and this is how the cohort can avoid getting trapped in local minima. CI differs from ACO as the autocatalytic nature of the ants is replaced by competitive nature of the cohorts. Instead of having tendency to follow most followed behavior, candidates in CI try to incorporate the best behavior in every iteration. This prevents the algorithm from getting caught into local minima by not relying on the behavior that is locally optimal. The CI algorithm has shown itself to be comparable to the best results obtained from the various techniques. The sharing of the best solution among candidates in CI gives a direction for all the candidates to move towards, but the independent search of each candidate ensures that the candidates come out of local minima to get the best solutions.

2.3 Validation of Cohort Intelligence

The performance of the proposed CI algorithm was tested by solving unconstrained well known test problems such as Ackley, Dixon and Price, Griewank, Hartmann, Levy, Michalewicz, Perm, Powell, Powersum, Rastrigin, Schwefel, Sphere, Sum Square, Trid, Zakhrov, etc. with different problem sizes. The algorithm was coded in MATLAB 7.8.0 (R2009A) on Windows platform using Intel Core 2 Duo T6570, 2.10 GHz processor speed and 2 GB RAM. Every test problem was solved 20 times with number of candidates C, number of variations in the behavior t chosen as 5 and 7, respectively. The values of the reduction factor r chosen for

Table 2.1 Summary of unconstrained test problem solutions with 5 variables

Problem	True optimum	CI algorithm Best Mean Worst	Function evaluations	Standard deviation	Time (s)
Ackley	0.0000	2.04E−10 5.58E−09 2.70E−08	43,493	1.06E−08	1.42
Dixon and Price	0.0000	9.86E−32 1.43E−09 4.28E−09	82,770	4.28E−09	1.83
Griewank	0.0000	0.00E+00 1.80E−02 3.70E−02	163,485	9.73E−03	5.15
Levy	0.0000	1.28E−21 6.22E−20 2.18E−19	38,993	7.70E−20	1.43
Michalewicz	−4.687658	−3.96E+00 −3.40E+00 −2.80E+00	188,700	3.61E−01	5.61
Perm	0.0000	4.20E−01 2.82E+00 9.83E+00	225,960	2.79E+00	8.07
Powell	0.0000	8.73E−09 2.31E−06 1.88E−05	162,510	5.54E−06	5.65
Rastrigin	0.0000	9.95E−01 1.50E+00 2.00E+00	277,440	4.91E−01	5.99
Schwefel	0.0000	2.83E−06 6.03E−06 8.86E−06	1,570,718	1.70E−06	38.73
Sphere	0.0000	2.69E−29 1.58E−28 2.55E−28	12,345	5.84E−29	0.35
Sum Square	0.0000	1.56E−18 2.96E−18 6.70E−18	7470	1.37E−18	0.22
Zakhrov	0.0000	8.38E−19 1.95E−18 3.20E−18	7470	6.62E−19	0.23

every unconstrained test problem are listed in Table 2.6. These parameters were derived empirically over numerous experiments.

The CI performance solving a variety of unconstrained test problems is presented in Tables 2.1, 2.2, 2.3, 2.4 and 2.5 with increase in the number of variables

Table 2.2 Summary of unconstrained test problem solutions with 10 variables

Problem	True optimum	CI algorithm Best Mean Worst	Function evaluations	Standard deviation	Time (s)
Ackley	0.0000	8.97E−08 4.58E−07 9.30E−07	30,765	3.60E−07	1.39
Dixon and Price	0.0000	6.67E−01 6.67E−01 6.67E−01	359,640	1.48E−14	12.11
Griewank	0.0000	7.48E−03 2.50E−02 4.68E−02	432,368	1.18E−02	17.49
Levy	0.0000	2.02E−06 7.34E−06 1.12E−05	44,798	2.78E−06	2.29
Powell	0.0000	7.73E−06 6.28E−05 2.13E−04	186,570	7.29E−05	9.46
Rastrigin	0.0000	6.96E+00 1.06E+01 1.49E+01	261,998	2.31E+00	10.04
Rosenbrock	0.0000	0.0000E+00 0.0000E+00 0.0000E+00	13,605	0.0000E+00	0.49
Schwefel	0.0000	1.20E−06 1.52E−06 1.74E−06	2,023,103	1.66E−07	84.55
Sphere	0.0000	7.17E−22 9.47E−22 1.62E−21	18,668	2.73E−22	0.76
Sum Square	0.0000	1.32E−16 2.37E−15 2.21E−14	14,948	6.58E−15	0.61
Zakhrov	0.0000	3.22E−12 5.15E−12 6.68E−12	22,365	1.24E−12	0.90

associated with the individual problem. It is observed that with increase in number of variables, the computational cost, i.e. function evaluations and computational time was increased. However, the small standard deviation values for all the

Table 2.3 Summary of unconstrained test problem solutions with 20 variables

Problem	True optimum	CI algorithm Best Mean Worst	Function evaluations	Standard deviation	Time (s)
Ackley	0.0000	2.97E−11 6.04E−11 2.96E−10	329,745	7.88E−11	21.87
Dixon and Price	0.0000	7.47E−01 7.77E−01 8.22E−01	358,800	2.31E−02	20.65
Griewank	0.0000	0.00E+00 7.40E−04 7.40E−03	187,763	2.22E−03	14.00
Levy	0.0000	7.31E−15 1.73E−13 8.79E−13	329,768	3.28E−13	27.01
Powell	0.0000	8.46E−05 2.18E−04 3.60E−04	539,640	8.02E−05	44.86
Rastrigin	0.0000	2.19E+01 3.88E+01 5.57E+01	408,758	7.87E+00	24.60
Rosenbrock	0.0000	0.00E+00 3.96E−30 1.98E−29	17,288	7.92E−30	1.01
Schwefel	0.0000	4.38E−06 5.56E−06 6.12E−06	2,023,950	5.03E−07	134.37
Sphere	0.0000	6.22E−14 7.60E−14 9.40E−14	26,183	1.11E−14	1.78
Sum Square	0.0000	8.88E−11 3.78E−10 1.88E−09	37,335	5.17E−10	2.50
Zakhrov	0.0000	1.00E−06 2.19E−06 4.97E−06	37,290	1.08E−06	2.53

functions independent of the number of variables highlighted its robustness. The effect of CI parameters such as number of candidates C, reduction rate r and number of variations in behavior t was also analyzed on unimodal as well as multimodal

Table 2.4 Summary of unconstrained test problem solutions with 30 variables

Problem	True optimum	CI algorithm Best Mean Worst	Function evaluations	Standard deviation	Time (s)
Ackley	0.0000	1.59E−07 5.91E−07 1.79E−06	299,835	6.41E−07	28.91
Dixon and Price	0.0000	9.89E−01 1.10E+00 1.23E+00	357,413	7.25E−02	29.07
Griewank	0.0000	2.90E−06 1.82E−03 7.57E−03	398,918	2.86E−03	39.91
Levy	0.0000	4.89E−07 4.62E−05 2.47E−04	674,363	9.15E−05	80.24
Powell	0.0000	7.23E−02 1.84E−01 4.23E−01	743,595	1.23E−01	89.86
Rastrigin	0.0000	5.57E+01 8.77E+01 1.00E+02	296,745	1.41E+01	26.40
Rosenbrock	0.0000	0.00E+00 7.22E−30 3.61E−29	19,470	1.44E−29	1.58
Schwefel	0.0000	1.06E−05 1.20E−05 1.37E−05	2,023,290	8.74E−07	180.62
Sphere	0.0000	1.91E−13 2.66E−13 4.51E−13	26,235	7.38E−14	2.51
Sum Square	0.0000	1.28E−10 2.52E−04 9.75E−04	55,433	3.79E−04	5.23
Zakhrov	0.0000	1.63E−04 6.93E−04 1.15E−03	221,798	3.03E−04	20.98

functions. The effect is visible in Fig. 2.3 where effect of these parameters on Sphere function and Ackley Function is presented as a representative to unimodal and multimodal function, respectively. The visualization for the convergence of the

Table 2.5 Summary of solution to Powersum, Hartmann and Trid function

Problem	No of variables	True optimum	CI algorithm Best Mean Worst	Function evaluations	Standard deviation	Time (s)
Powersum	4	0.0000	1.38E−06 6.74E−05 1.34E−04	619,125	4.44E-05	17.48
Hartmann	6	−3.86278	−3.32E+00 −3.32E+00 −3.32E+00	710,483	1.67E-03	31.53
Trid	6	−50.0000	−4.87E+01 −4.87E+01 −4.87E+01	177,090	4.92E-03	4.27

representative Ackley function is presented in Fig. 2.2 for learning attempts 1, 10, 15 and 30. For both types of functions, the computational cost, i.e. function evaluations and computational time was observed to be increasing linearly with increasing number of candidates C (refer to Fig. 2.3a, b) as well as number of variations in behavior t (refer to Fig. 2.3e, f, k and l). This was because, with increase in number of candidates, number of behavior choices i.e. function evaluations also increased. Moreover, with fewer number of candidates C, the quality of the solution at the end of first learning attempt referred to as initial solution as well as the converged solution were quite close to each other and importantly, the converged solutions and the converged solution was suboptimal. The quality of both the solutions improved with increase in number of candidates C. This was because fewer number of behavior choices were available with fewer number of candidates, whereas with increase in number of candidates the total choice of behavior also increased as a result initial solution worsened whereas converged solution improved as sufficient time was provided for saturation. Due to this a widening gap between initial solution and converged solution was observed in the

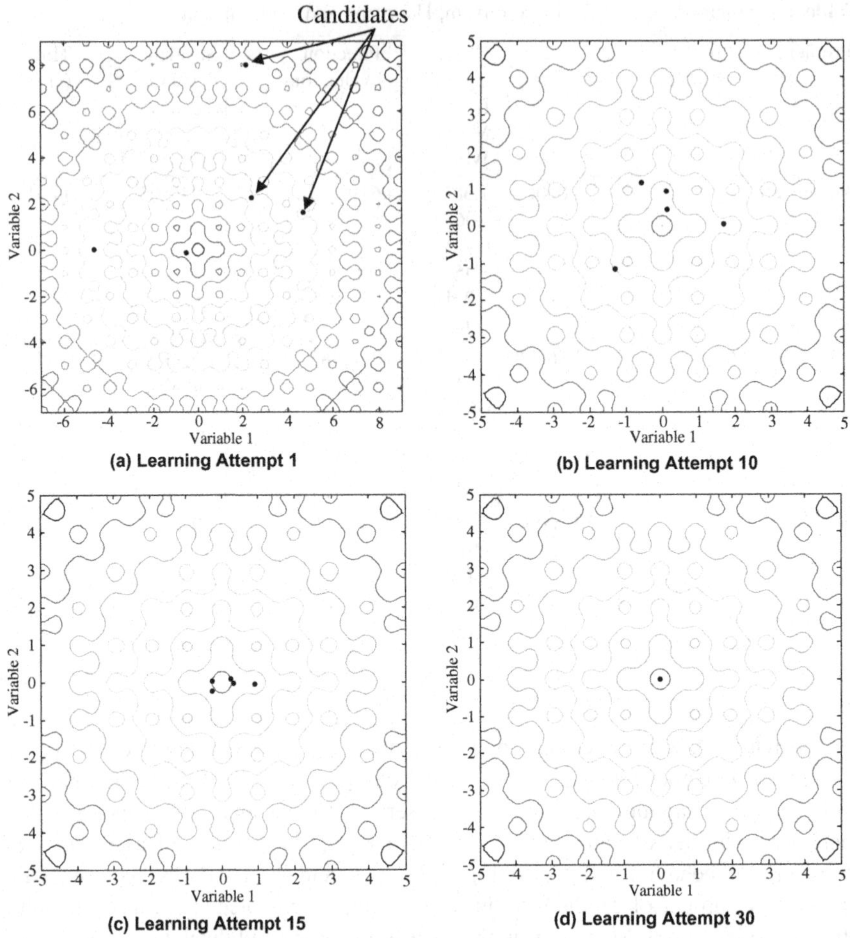

Fig. 2.2 Visualization of convergence of solutions at various learning attempts for Ackley function

trend of initial solution and final solution. It is evident in Fig. 2.3g, h. In addition, it is important observation from Table 2.6 that larger values of the reduction rate *r* were required as the problem size increased. This could be because as the number of variables increase size of the problem search space also increased and larger values of the reduction rate *r* were required. However, it is evident from Fig. 2.3c, d, i, j that for the same size of unimodal as well as multimodal problems with the

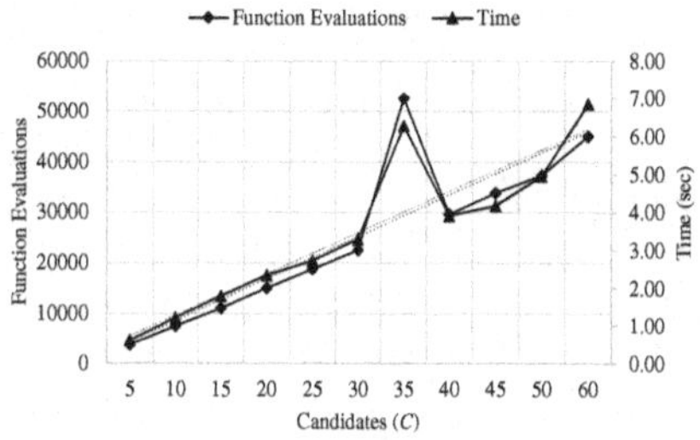

(a) Variation in Function Evaluations and Time with respect to Number of Candidates (*C*)

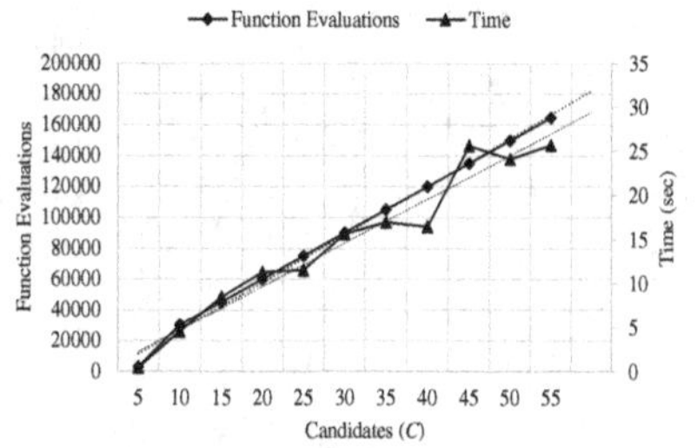

(b) Variation in Function Evaluations and Time with respect to Number of Candidates (*C*)

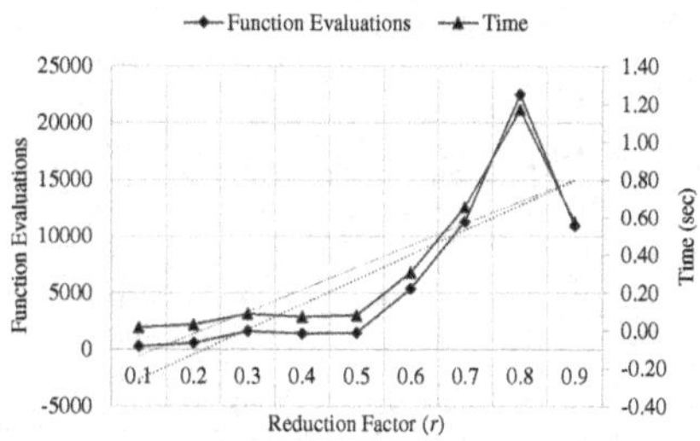

(c) Variation in Function Evaluations and Time with respect to Reduction Factor (*r*)

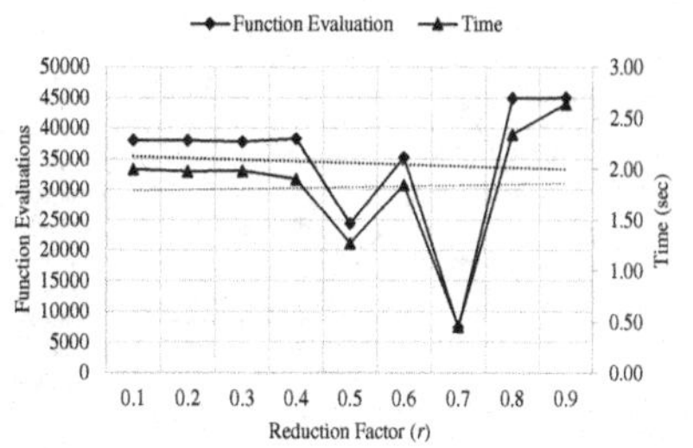

(d) Variation in Function Evaluations and Time with respect to Reduction Factor (*r*)

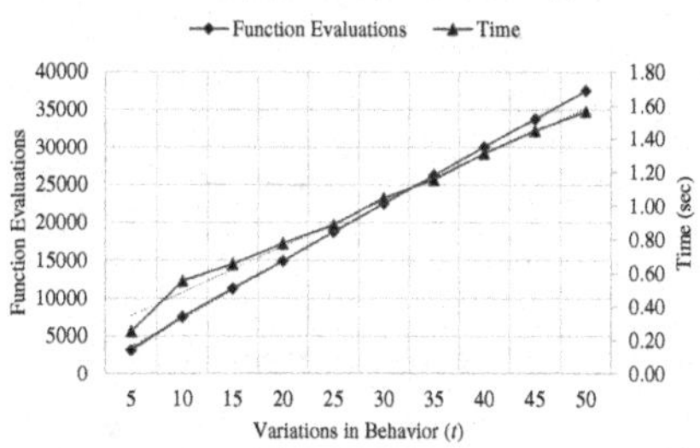

(e) Variation in Function Evaluations and Time with respect to Variation in Behavior (*t*)

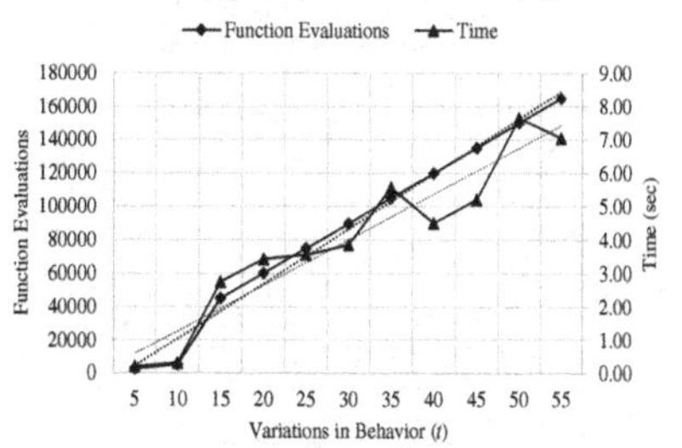

(f) Variation in Function Evaluations and Time with respect to Variation in Behavior (*t*)

Fig. 2.3 Influence of number of candidates (*C*), reduction factor (*r*) and variation in behavior (*t*) on CI algorithm performance

increase in the value of reduction rate *r* larger solution space was available for exploration which when searched resulted into increased converged solution quality as well as associated computational cost. In addition, similar to the increase in number of candidates *C*, it was observed that with the increase in number of variations in behavior *t* the computational cost increased along with the

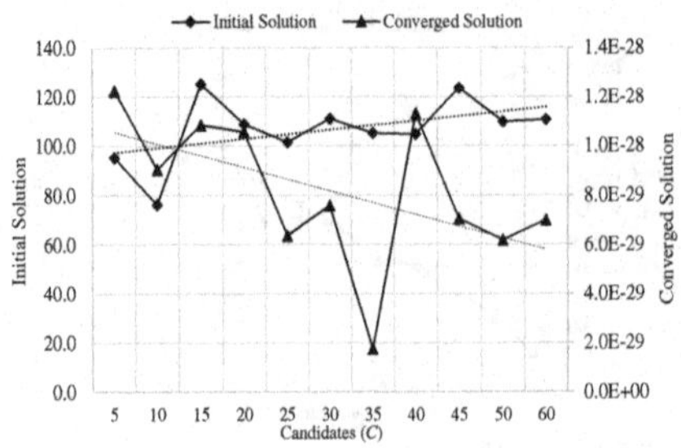

(g) Variation in Initial and Final Solution with respect to Number of Candidates (*C*)

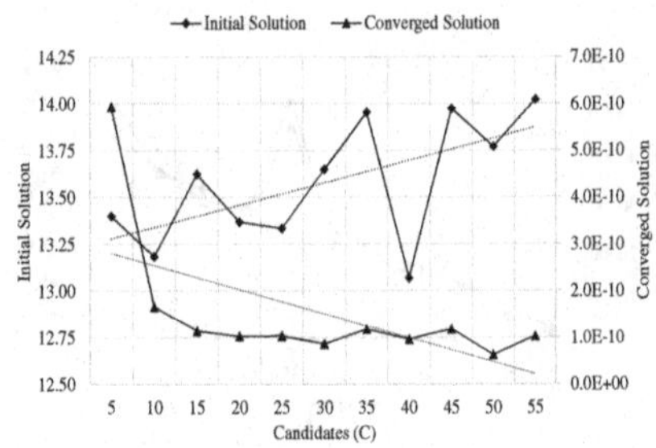

(h) Variation in Initial and Final Solution with respect to Number of Candidates (*C*)

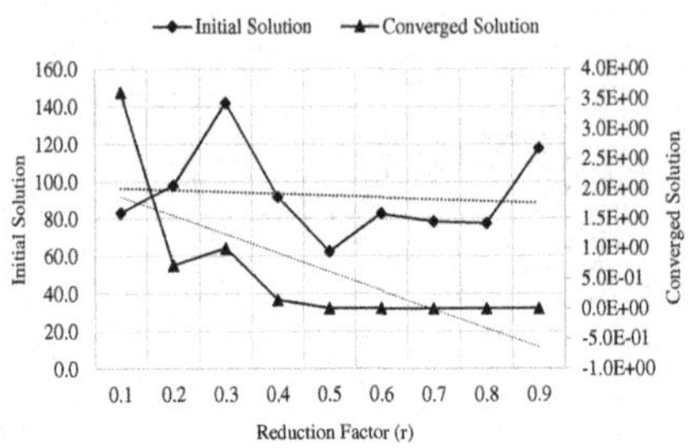

(i) Variation in Initial and Final Solution with respect to Reduction Factor (*r*)

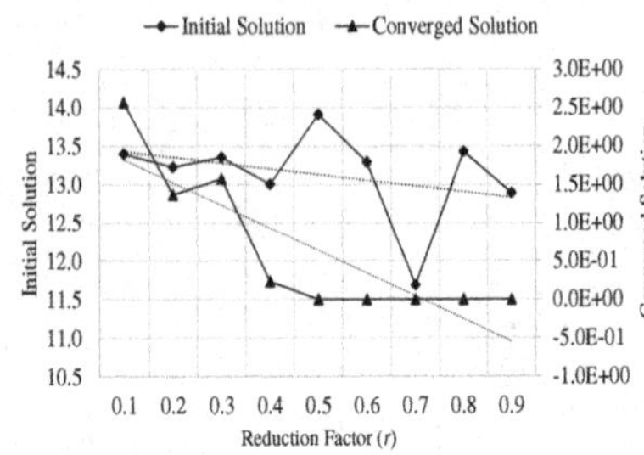

(j) Variation in Initial and final Solution with respect to Reduction Factor (*r*)

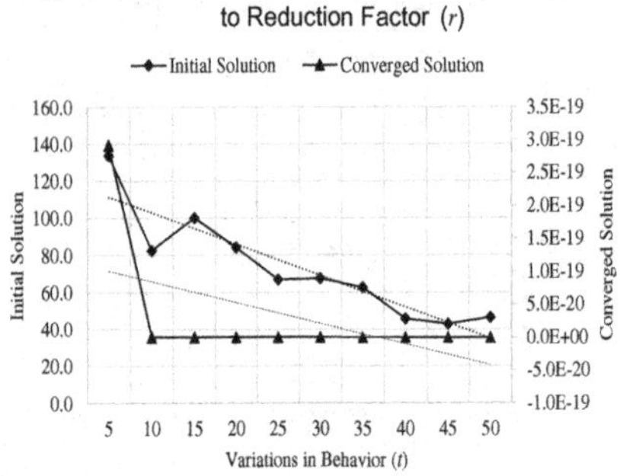

(k) Variation in Initial and Final Solution with respect to Variation in Behavior (*t*)

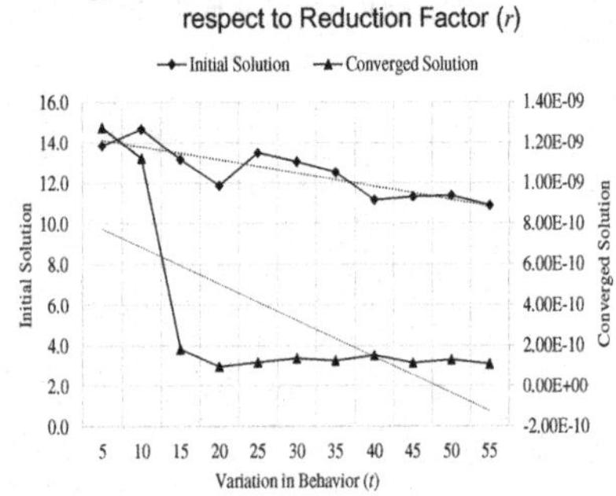

(l) Variation in Initial and Final Solution with respect to Variation in Behavior (*t*)

Fig. 2.3 (continued)

improvement in converged solution quality. Furthermore, the converged solution quality did not improve significantly after a certain number of variations in behavior *t*. Moreover, with the increase in variations in behavior *t*, the difference between the initial solution and the converged solution gradually decreased.

Table 2.6 Summary of reduction factor r values for unconstrained test problems

Problem	Reduction factor r					
	$n = 4$	$n = 5$	$n = 6$	$n = 10$	$n = 20$	$n = 30$
Ackley	–	0.800	–	0.850	0.970	0.980
Dixon and Price	–	0.850	–	0.980	0.997	0.997
Griewank	–	0.950	–	0.997	0.950	0.980
Hartmann	–	–	0.997	–	–	–
Levy	–	0.800	–	0.950	0.980	0.995
Michalewicz	–	0.950	–	0.970	0.980	0.970
Perm	–	0.997	–	–	–	–
Powell	–	0.850	–	0.990	0.950	0.980
Powersum	0.997	–	–	–	–	–
Rastrigin	–	0.980	–	0.900	0.980	0.980
Rosenbrock	–	–	–	–	–	–
Schwefel	–	0.997	–	0.997	0.997	0.997
Sphere	–	0.800	–	0.900	0.950	0.950
Sum Square	–	0.800	–	0.900	0.970	0.980
Trid	–	–	0.800	0.980	0.980	0.980
Zakhrov	–	0.800	–	0.950	0.980	0.980

References

1. Deb, K.: An efficient constraint handling method for genetic algorithms. Comput. Methods Appl. Mech. Eng. **186**, 311–338 (2000)
2. Ray, T., Tai, K., Seow, K.C.: Multiobjective design optimization by an evolutionary algorithm. Eng. Optim. **33**(4), 399–424 (2001)
3. Kennedy, J., Eberhart, R.: Particle swarm optimization. In: Proceedings of IEEE International Conference on Neural Networks, pp. 1942–1948 (1995)
4. Dorigo, M., Birattari, M., Stitzle, T.: Ant colony optimization: arificial ants as a computational intelligence technique. IEEE Comput. Intell. Mag., 28–39 (2006)
5. Pham, D.T., Ghanbarzadeh, A., Koc, E., Otri, S., Rahim, S., Zaidi, M.: The bees algorithm. Technical Note, Manufacturing Engineering Centre, Cardiff University, UK (2005)
6. Kulkarni, A.J., Durugkar, I.P, Kumar, M.R.: Cohort intelligence: a self-supervised learning behavior. In: Proceedings of the IEEE Conference on Systems, Man and Cybernetics 2013, pp. 1396–1400 (2013)
7. Kulkarni, A.J., Shabir, H.: Solving 0-1 Knapsack problem using cohort intelligence algorithm. Int. J. Mach. Learn. Cybernet. (2014). doi:10.1007/s13042-014-0272-y
8. Wang, H.: Opposition-based barebones particle swarm for constrained nonlinear optimization problems. Math. Probl. Eng. **2012**, Article ID 761708 (2010)
9. Coelho, L.D.S.: Gaussian quantum-behaved particle swarm optimization approaches for constrained engineering design problems. Expert Syst. Appl. **37**(2), 1676–1683 (2010)
10. Sun, C.L., Zeng, J.C., Pan, J.S.: An new vector particle swarm optimization for constrained optimization problems. In: Proceedings of the International Joint Conference on Computational Sciences and Optimization (CSO'09), pp. 485–488 (2009)
11. Kennedy, J.: Bare bones particle swarms. In: Proceedings of the IEEE Swarm Intelligence Symposium (SIS'03), pp. 80–87 (2003)

12. Mendes, R., Kennedy, J., Neves, J.: The fully informed particle swarm: simpler, maybe better. IEEE Trans. Evolut. Comput. **8**(3), 204–210 (2004)
13. Chen, L., Sun, H., Wang, S.: Solving continuous optimization using ant colony algorithm. In: Second International Conference on Future Information Technology and Management Engineering, pp. 92–95 (2009)
14. Dorigo, M., Maniezzo, V., Colorni, A.: Ant system: optimization by a colony of cooperating agents. IEEE Trans. Syst. Man Cybernet Part B: Cybernet **26**(1), 29–41 (1996)

Chapter 3
Cohort Intelligence for Constrained Test Problems

Any optimization algorithm requires a technique/way to handle constraints. This is important because most of the real world problems are inherently constrained problems. There are a several traditional methods available such as feasibility-based methods, gradient projection method, reduced gradient method, Lagrange multiplier method, aggregate constraint method, feasible direction based method, penalty based method, etc. [1]. According to Vanderplaat [2], the penalty based methods can be referred to as generalized constraint handling methods. They can be easily incorporated into most of the unconstrained optimization methods and can be used to handle nonlinear constraints. Another approach is feasibility-based approach. Similar to the penalty based methods, it is also simple to use as it assists the unconstrained optimization methods to drive into feasible region and further reach in close neighborhood of the optimum solution [3–5]. Similar to other nature-/bio-inspired techniques, the performance of Cohort Intelligence (CI) methodology may degenerate when applied for solving constrained problems. As an effort in the direction of developing and further incorporating a generic constraint handling technique into the CI framework, a penalty function approach is used. The performance of the constrained CI approach is tested by successfully solving several well studied constrained test problems.

3.1 Constraint Handling Using Penalty Function Approach

Consider a general constrained problem (in the minimization sense) as follows:

$$\begin{aligned} &\text{Minimize} && f(\mathbf{x}) \\ &\text{Subject to} && g(\mathbf{x})_j \le 0, \quad j = 1, 2, \ldots, s \\ & && h(\mathbf{x})_j = 0, \quad j = 1, 2, \ldots, w \\ &\text{Subject to} && \Psi_i^{lower} \le x_i \le \Psi_i^{upper}, \quad i = 1, \ldots, N \end{aligned} \tag{3.1}$$

A.J. Kulkarni et al., *Cohort Intelligence: A Socio-inspired Optimization Method*, Intelligent Systems Reference Library 114, DOI 10.1007/978-3-319-44254-9_3

In order to incorporate the constraints into the problem, a pseudo-objective function is formed as follows:

$$\phi(\mathbf{x}) = f(\mathbf{x}) + \theta\left\{\sum_{j=1}^{s}\left[g_j^+(\mathbf{x})\right]^2 + \sum_{j=1}^{w}\left[h_j(\mathbf{x})\right]^2\right\} \tag{3.2}$$
$$\text{Subject to}\quad \Psi_i^{lower} \le x_i \le \Psi_i^{upper},\quad i = 1,\ldots,N$$

where $g_j^+(\mathbf{x}) = \max(0, g_j(\mathbf{x}))$ and θ is the scalar penalty parameter which is fixed in all the runs of the CI algorithm.

3.2 Numerical Experiments and Discussion

CI methodology was applied to a variety of well known constrained test problems with penalty function approach [1–5] incorporated into it. The CI algorithm was coded in MATLAB 7.14.0.739 (R2012a) on Windows platform using Intel Core 2 Duo, 1.67 GHz processor speed and 2 GB RAM. Every problem was solved 20 times with number of candidates C chosen as 5 and number of variations in the behavior t chosen for constrained test problem is listed in Table 3.9. These parameters were derived empirically over numerous experiments.

It is evident from the results presented in Tables 3.1, 3.2, 3.3, 3.4, 3.5, 3.6, 3.7, 3.8 and 3.9, that the CI methodology is capable of efficiently handling a variety of equality as well as inequality of constraints. For detailed properties of these problems the reader is encouraged to refer to [4]. The results also demonstrated the competitiveness with other contemporary methods. Furthermore, the standard deviation presented in Table 3.9, it is evident that the approach was sufficiently robust with reasonable computational cost, i.e. function evaluations and computational time. According to Table 3.9, the solutions obtained using CI were quite close to the best reported solution so far. As mentioned before, the parameters such as number of candidates C, number of variations in the behavior t and reduction factor r were chosen empirically

Table 3.1 Characteristic of benchmark problems [4]

Problem	DV	Form of $f(\mathbf{x})$	(x%)	LI	NE	NI	α
G03	10	Polynomial	0.002	0	1	0	1
G04	5	Quadratic	52.123	0	0	6	2
G05	4	Cubic	0.000	2	3	0	3
G06	2	Cubic	0.006	0	0	2	2
G07	10	Quadratic	0.000	3	0	5	6
G08	2	Nonlinear	0.856	0	0	2	0
G09	7	Polynomial	0.512	0	0	4	2
G11	2	Quadratic	0.000	0	1	0	1

Table 3.2 Summary of the constrained test problem solutions

Problem		G03	G04	G05	G06	G07	G08	G09	G11
Best known solutions		1	−30,665.539	5126.498	−6961.81388	24.306	0.095825	680.6300573	0.75
Methods	Koziel et al. (1999) [23]	Best 0.9997 Avg. 0.9989 Worst 0.9960	Best −30664.5 Avg. −30655.3 Worst −30645.9	N.A. N.A. N.A.	Best −6952.1 Avg. −6342.6 Worst −5473.9	Best 24.620 Avg. 24.826 Worst 24.069	Best 0.0958250 Avg. 0.0891568 Worst 0.0291438	Best 680.91 Avg. 681.16 Worst 683.18	Best 0.75 Avg. 0.75 Worst 0.75
	Runarsson et al. (2000) [21]	1.000 1.000 1.000	−30,665.539 −30,665.539 −30,665.539	5126.497 5128.881 5142.472	−6961.814 −6875.940 −6350.262	24.307 24.374 24.642	0.095825 0.095825 0.095825	680.630 680.656 680.763	0.750 0.750 0.750
	Hamida et al. (2002) [22]	1 0.99989 N.A.	−30,665.5 −30,665.5 N.A.	5126.5 5141.65 N.A.	−6961.81 −6961.81 N.A.	24.3323 24.6636 N.A	0.09582 0.09582 N.A.	680.630 680.641 N.A.	0.75 0.75 N.A.
	Montes et al. (2003) [23]	1.001038 1.000989 1.000579	−30,665.539062 −30,665.539062 −30,665.539062	5126.599609 5174.492301 5304.166992	−6961.813965 −6961.283984 −6961.481934	24.326715 24.474926 24.842829	0.095826 0.095826 0.095826	680.631592 680.643410 680.719299	0.749090 0.749358 0.749830
	Hedar et al. (2006) [15]	1.0000015 0.9991874 0.9915186	−30,665.5380 −30,665.4665 −30,664.6880	5126.4981 5126.4981 5126.4981	−6961.81388 −6961.81388 −6961.81388	24.310571 24.379527 24.644397	0.095825 0.095825 0.095825	680.63008 680.63642 680.69832	0.7499990 0.7499990 0.7499990
	Bacerra et al. (2006) [11]	0.995143 0.788635 0.639920	−30,665.538672 −30,665.538672 −30,665.538672	5126.5709 5207.4106 5327.3904	−6961.813876 −6961.813876 −6961.813876	24.306209 24.306210 24.306212	0.095825 0.095825 0.095825	680.630057 680.630057 680.630057	0.749900 0.757995 0.796455
	Lampinen (2006) [6]	N.A. N.A. N.A.	N.A. N.A. N.A.	5126.484 5126.484 5126.484	−6961.814 −6961.814 −6961.814	N.A. N.A. N.A.	0.095825 0.095825 0.095825	680.630 680.630 680.630	0.74900 0.74900 0.74900
	Chootinan et al. (2006) [12]	0.99998 0.99979	−30,665.5386 −30,665.5386 −30,665.5386	5126.4981 5126.4981 5126.4981	−6961.8139 −6961.8139 −6961.8139	N.A. N.A. N.A.	0.0958250 0.0958250 0.0958250	680.6303 680.6381 680.6538	0.7500 0.7500 0.7500

(continued)

Table 3.2 (continued)

Problem		G03	G04	G05	G06	G07	G08	G09	G11
	Farmani et al. (2003) [4]	0.99978	−30,665.5000	5126.9890	−6961.8000	24.59	0.0958250	680.6400	0.7500
		0.99930	−30,665.2000	5432.08	−6961.8000	27.83	0.0958250	680.7200	0.7500
		0.99830	−30,663.3000	N.A.	−6961.8000	32.69	0.0958250	680.8700	0.7500
	Deb (2000) [5]	N.A.	−30,614.814	N.A.	N.A.	24.372	N.A.	680.659424	N.A.
		N.A.	−30,196.404	N.A.	N.A.	24.409	N.A.	681.525635	N.A.
		N.A.	−29,606.596	N.A.	N.A.	25.075	N.A.	687.188599	N.A.
	Dong et al. (2007) [14]	N.A.	−30,664.7	N.A.	N.A.	N.A.	N.A.	N.A.	N.A.
		N.A.	−30,656.1	N.A.	N.A.	N.A.	N.A.	N.A.	N.A.
		N.A.	−30,662.8	N.A.	N.A.	N.A.	N.A.	N.A.	N.A.
	He et al. (2007) [7]	N.A.	−30,665.539	N.A.	N.A.	N.A.	0.095825	N.A.	N.A.
		N.A.	−30,665.539	N.A.	N.A.	N.A.	0.095825	N.A.	N.A.
		N.A.	−30,665.539	N.A.	N.A.	N.A.	0.095825	N.A.	N.A.
	Hu et al. (2002) [8]	1	−30,665.5	N.A.	−6961.7	24.4420	0.09583	680.657	0.7500
		1	−30,665.5	N.A.	−6960.7	26.7188	0.09583	680.876	0.7500
		1	−30,665.5	N.A.	−6956.8	31.1843	0.09583	681.675	0.7500
	Ray et al. [24]	N.A.	−30,651.662	N.A.	N.A.	N.A.	N.A.	N.A.	N.A.
		N.A.	−30,647.105	N.A.	N.A.	N.A.	N.A.	N.A.	N.A.
		N.A.	−30,619.047	N.A.	N.A.	N.A.	N.A.	N.A.	N.A.
	CI algorithm	0.998892	−30,665.531736	5143.533669	−6961.812948	24.310498	0.095825	680.731701	0.749904
		0.999417	−30,665.529486	5196.042840	−6961.777472	24.357417	0.095825	680.921574	0.752580
		0.999762	−30,665.526082	5273.835265	−6961.569046	24.403683	0.095825	681.226784	0.760863

Table 3.3 Performance comparison of various algorithms solving spring design problem

Design variables	Best solutions found							
	Arora (2004) [25]	Coello (2000) [3]	Coello et al. (2002) [16]	Coello et al. (2004) [10]	He et al. (2006) [26]	He et al. (2007) [7]	Kulkarni et al. (2011) [1]	CI algorithm
x_1	0.05339	0.05148	0.05198	0.05000	0.051728	0.05170	0.05060	0.050978
x_2	0.39918	0.35166	0.36396	0.31739	0.357644	0.35712	0.32781	0.339856
x_3	9.18540	11.63220	10.89052	14.03179	11.244543	11.26508	14.05670	12.350557
$g_1(\mathbf{X})$	0.00001	−0.00330	−0.000013	0.00000	−0.000845	−0.00000	−0.05290	2.6906e −007
$g_2(\mathbf{X})$	−0.00001	−0.00010	−0.000021	−0.00007	−1.2600e−05	0.00000	−0.00740	−8.2557e −007
$g_3(\mathbf{X})$	−4.12383	−4.02630	−4.061338	−3.96796	−4.051300	−4.05460	−3.70440	−4.0191
$g_4(\mathbf{X})$	−0.69828	−0.73120	−0.722698	−0.75507	−0.727090	−0.7274	−0.74769	−0.7394
$f(\mathbf{X})$	0.01273	0.01270	0.01268	0.01272	0.0126747	0.01266	0.01350	0.012675

Table 3.4 Statistical results of different methods solving spring design problem

Methods	Best	Mean	Worst	Std.
Arora (2004) [25]	0.0127303	N.A.	N.A.	N.A.
Coello (2000) [3]	0.0127048	0.0127690	0.012822	3.9390e−005
Coello et al. (2002) [16]	0.0126810	0.0127420	0.012973	5.9000e−005
Coello et al. (2004) [10]	0.0127210	0.0135681	0.015116	8.4152e−004
He et al. (2006) [26]	0.0126747	0.0127300	0.012924	5.1985e−004
He et al. (2007) [7]	0.0126652	0.0127072	0.0127191	1.5824e−005
Kulkarni et al. (2011) [1]	0.01350	0.02607	0.05270	N.A.
CI Algorithm	0.012679	0.012719	0.012884	0.000062

over numerous experiments. Since the chosen set of parameters produced sufficiently robust results much effort was not spent in their fine-tuning. Hence, better performance may be obtained through different choice of parameters.

The CI was incorporated with penalty function approach [1, 2] and tested by solving several well studied constrained problems. The results were compared with existing contemporary approaches. A few approaches focused on overcoming the limitation of penalty approach. A self-adaptive penalty approach [3], a dynamic penalty scheme [4], GA with binary representation assisted with traditional penalty function method [5], etc. resulted in premature convergence with high sensitivity to additional parameters. PSO uses penalty factors as searching variables but it is weak in local searches. Also it is not efficient in maintaining balance between exploitation and exploration due to lack of diversity. The approach of penalty parameter was avoided by utilizing feasibility based rule in [5] solving constrained problems. It failed to produce optimum solution in every run and also required an extra fitness function. A variation of feasibility based rule [5] was proposed in [6] for solving constrained non-linear functions. In both the approaches the population in hand is the governing factor of the quality of solutions. The need for an extra fitness function was avoided by HPSO [7] by introducing the feasibility based rule [5] into PSO. The PSO [8] and the homomorphous mapping [9] required feasible solution initially along with set dependent parameters. For some problems, it is quite hard to generate feasible solutions initially and requires additional techniques. In cultural algorithm [10] and cultural differential evolution (CDE) [11] it is seen that there is a lack in diversity of the population. The gradient repair method [12] was implanted into PSO [13] and the number of solution undergoing repair [14] are the key factors of its performance. Taking directions from [15], GA was applied to find solution vector (non-dominated) [16]. In addition, Genetics Adaptive Search (Gene AS) [17], augment Lagrange multiplier method [18], geometric programming approach [19], and a branch and bound technique [20] were also used for solving various constrained benchmark problems addressed above which required additional gradient method.

Table 3.5 Performance comparison of various Algorithms solving Welded Beam Design Problem

Design var.	Coello (2000) [3]	Coello et al. (2002) [16]	Coello et al. (2004) [10]	He et al. (2006) [26]	He et al. (2007) [7]	Deb (2000) [5]	Siddall (1972) [27]	Ragsdell et al. (1976) [19]	CI algorithm
x_1	0.208800	0.205986	0.205700	0.202369	0.20573	0.2489	0.2444	0.2455	0.205730
x_2	3.420500	3.471328	3.470500	3.544214	3.47048	6.1730	6.2189	6.1960	3.635788
x_3	8.997500	9.020224	9.036600	9.048210	9.03662	8.1789	8.2915	8.2730	9.036623
x_4	0.210000	0.206480	0.205700	0.205723	0.20573	−0.2533	0.2444	0.2455	0.205729
$g_1(\mathbf{X})$	−0.337812	−0.074092	−0.00047	−12.83979	0.00010	−5758.6037	−5743.50202	−5743.82651	−0.0115
$g_2(\mathbf{X})$	−353.90260	−0.266227	−0.00156	−1.247467	−0.02656	−255.57690	−4.015209	−4.715097	−7.5194e −004
$g_3(\mathbf{X})$	−0.00120	−0.000495	0.000000	−0.001498	0.00000	−0.004400	0.000000	0.000000	−8.5103e −008
$g_4(\mathbf{X})$	−3.411865	−3.430043	−3.43298	−3.429347	−3.43298	−2.982866	−3.022561	−3.020289	−3.4182
$g_5(\mathbf{X})$	−0.08380	−0.080986	−0.08073	−0.079381	−0.08070	−0.123900	−0.119400	−0.120500	−0.0807
$g_6(\mathbf{X})$	−0.235649	−0.235514	−0.23554	−0.235536	−0.23554	−0.234160	−0.234243	−0.234208	−0.2355
$g_7(\mathbf{X})$	−363.23238	−58.66644	−0.00077	−11.68135	−0.02980	−4465.27092	−3490.46941	−3604.27500	−0.0021
$f(\mathbf{X})$	1.748309	1.728226	1.724852	1.728024	1.72485	2.43311600	2.38154338	2.38593732	1.770316

Table 3.6 Statistical results of different methods solving welded beam design problem

Methods	Best	Mean	Worst	Std.
Coello (2000) [3]	1.748309	1.771973	1.785835	0.011220
Coello et al. (2002) [16]	1.728226	1.792654	1.993408	0.074713
Coello et al. (2004) [10]	1.724852	1.971809	3.179709	0.443131
He et al. (2006) [26]	1.728024	1.748831	1.782143	0.012926
He et al. (2007) [7]	1.724852	1.749040	1.814295	0.040049
Deb (2000) [5]	2.38145	2.38263	2.38355	N.A.
Siddall (1972) [27]	2.3815	N.A.	N.A.	N.A.
Ragsdell et al. (1976) [19]	2.3859	N.A.	N.A.	N.A.
CI algorithm	1.770436	1.779802	1.816707	0.013885

In this study, we apply the CI algorithm with penalty function approach to solve the benchmark problems studied by Coello and Becerra [10]. This test problems solved include two maximization problems (G03 and G08) and six minimization problems (G04, G05, G07, G08, G09, G11). Since all constraints have explicit and simple functional forms, the gradient of the constraints can be derived directly from the constraint set. The characteristics of problems, including the number of decision variables *(DV)*, form of objective function, size of feasible region ($\mathbf{x}\%$), and the number of linear inequality (*LI*), non-linear equality (*NE*), non-linear inequality (*NI*), and number of active constraints at the reported optimum (a), are summarized in Table 3.1. The size of feasible region, empirically determined by simulation, indicates the difficulty to randomly generate a feasible solution. These problems have been grouped in different categories [4] (refer to Table 3.1) based on the problem characteristics such as, nonlinear objective function (NLOF) [G03, G04, G05, G06, G07, G08, G09, G11], nonlinear equality constraints (NEC) (G03, G05, G11), moderated dimensionality (MD) ($n \geq 5$) (G03, G07, G09), active constraints (AC) ($a \geq 6$) (G07) and small feasible region (SFR) ($(\mathbf{x}\%) \approx 0$) (G03, G05, G11).

As presented in Tables 3.1, 3.2, 3.3, 3.4, 3.5, 3.6, 3.7, 3.8 and 3.9, the overall performance of CI with penalty function approach was quite comparable to other existing algorithms. The solution of to G03 problem was within 0.0583 % of the reported optimum [8, 21]. Also, the computational cost, [time and Function Evaluations (FEs)] was quite reasonable. The CI performance for solving the problem G04 was within 0.0000326 % of the reported optimum [11, 12, 21–23] with a reasonable computational cost. The CI solutions to problems G06, G08 and G09 were well within 0.0005242, 0.00000 and 0.01218 %, respectively of the reported optimum [11, 24] with reasonable computational cost and standard deviation. The CI algorithm could solve the problem G05 within 1.0356 % of the reported optimum [21] with standard deviation of 55.0329. Also, the average computational time (30 s) was comparatively higher as compared to solving other problems using CI. This underscored that CI needs to be further modified to make it

Table 3.7 Performance comparison of various algorithms solving pressure vessel design problem

Des. var.	Sandgren (1988) [20]	Kannan et al. (1994) [18]	Deb (1997) [17]	Coello (2000) [3]	Coello et al. (2002) [16]	He et al. (2006) [26]	He et al. (2007) [7]	CI algorithm
x_1	1.1250	1.1250	0.9375	0.8125	0.8125	0.8125	0.81250	14
x_2	0.6250	0.6250	0.5000	0.4375	0.4375	0.4375	0.43750	7
x_3	47.7000	58.2910	48.3290	40.3239	42.0974	42.0913	42.09840	45.3368
x_3	117.7010	43.6900	112.6790	200.0000	176.6540	176.7465	176.6366	140.2538
$g_1(\mathbf{X})$	−0.204390	0.000016	−0.004750	−0.034324	−0.000020	−0.000139	-8.8×10^{-7}	−1.7306e −007
$g_2(\mathbf{X})$	−0.169942	−0.068904	−0.038941	−0.052847	−0.035891	−0.035949	−0.03590	−0.0050
$g_3(\mathbf{X})$	54.226012	−21.22010	−3652.87683	−27.10584	−27.886075	−116.38270	3.12270	−0.0280
$g_4(\mathbf{X})$	−122.29900	−196.31000	−127.32100	−40.00000	−63.34595	−63.253500	−63.36340	−99.7461
$f(\mathbf{X})$	8129.8000	7198.0428	6410.3811	6288.7445	6059.9463	6061.0777	6059.7143	6090.5000

Table 3.8 Statistical results of different methods for solving pressure vessel design problem

Methods	Best	Mean	Worst	Std.
Sandgren (1988) [20]	8129.8000	N.A.	N.A.	N.A.
Kannan et al. (1994) [18]	7198.0428	N.A.	N.A.	N.A.
Deb (1997) [17]	6410.3811	N.A.	N.A.	N.A.
Coello (2000) [3]	6288.7445	6293.8432	6308.1497	7.4133
Coello et al. (2002) [16]	6059.9463	6177.2533	6469.3220	130.9297
He et al. (2006) [26]	6061.0777	6147.1332	6363.8041	86.4545
He et al. (2007) [7]	6059.7143	6099.9323	6288.6770	86.2022
CI algorithm	6090.52639	6090.52689	6090.52849	0.00063

solve the problems with equality constraints. A mechanism similar to the window approach proposed by Ray et al. [24] could be devised. It is also important to mention that the value of reduction factor r varied from very narrow range of 0.98–0.998. This is in contrary to the CI solutions solving unconstrained test problems (refer to Chap. 2 for details). In addition, the solutions to problem G07 and G11 were within 0.211 and 0.344 %, respectively of the reported optimum [11, 12] also required higher computational cost.

Furthermore, the performance of CI was tested solving three well known problems from the mechanical engineering design domain. The approach of CI produced better results for solving the spring design problem within 0.11848 % of the reported optimum [7]. The solution was quite robust with standard deviation 0.000062. In addition, CI solution solving the welded beam design problem and the pressure vessel design problem yielded was within 2.6357 and 0.5049 %, respectively of the reported optimum [10, 16]. Also, the standard deviations and computational cost were quite reasonable.

3.3 Conclusions

The chapter has validated the constraint handling ability of the CI methodology by solving a variety of well known test problems. This also justified the possible application of CI for solving a variety of real world problems. In all the problem solutions, the implemented CI methodology produced sufficiently robust results with reasonable computational cost. It is important to mention here that similar to the original CI approach discussed in Chap. 2, the sampling space was restored to the original one when no significant improvement in the cohort behavior was observed. This helped the solution jump out of possible local minima.

In addition to the advantages few limitations are also observed. The computational performance was essentially governed by the parameter such as sampling

Table 3.9 CI solution details

Problem	Solutions Best Mean Worst	Standard deviation	Avg. no. of function evaluations	Avg. comp. time (seconds)	Closeness to the best reported solution (%)	Reduction factor r
G03	0.998892 0.999417 0.999762	0.000297	28,125	5.5	0.0583	0.98
G04	−30,665.531736 −30,665.529486 −30,665.526082	0.001740	25,125	5.8	0.0000326	0.99
G05	5143.533669 5196.042840 5273.835265	55.032911	42,375	30	1.035657	0.99
G06	−6961.812948 −6961.777472 −6961.569046	0.075814	27,335	4.5	0.0005242	0.98
G07	24.310498 24.357417 24.403683	0.029861	348,750	42.5	0.211540	0.998
G08	0.095825 0.095825 0.095825	0.000000	6000	1.59	0	0.98
G09	680.731701 680.921574 681.226784	0.177770	24,375	15.3	0.012184	0.99
G11	0.749904 0.752580 0.760863	0.004407	25,875	12.5	0.344	0.995
Spring design problem	0.012679 0.012719 0.012884	0.000062	313,500	58	0.11848	0.99
Welded beam design problem	1.770436 1.779802 1.816707	0.013885	25,000	165	2.6359	0.983
Pressure vessel design problem	6090.526390 6090.526895 6090.528495	0.000632	294,000	85	0.5049	0.9970

interval reduction factor r, number of candidates C and number of variations t. In the future, to make the approach more generalized and to improve the rate of convergence, the quality of results, as well as reduce the computational cost, a self adaptive scheme could be developed for these parameters. The authors also see strong potential in the field of game development and mutual learning.

References

1. Kulkarni, A.J., Tai, K.: Solving constrained optimization problems using probability collectives and a penalty function approach. Int. J. Comput. Intell. Appl. **10**(4), 445–470 (2011)
2. Vanderplaat, G.N.: Numerical Optimization Techniques for Engineering Design. McGraw-Hill, New York (1984)
3. Coello Coello, C.A.: Use of self-adaptive penalty approach for engineering optimization problems. Comput. Ind. **41**, 113–127 (2000)
4. Farmani, R., Wright, J.A.: Self-adaptive fitness formulation for constrained optimization. IEEE Trans. Evol. Comput. **7**(5), 445–455 (2003)
5. Deb, K.: An efficient constraint handling method for genetic algorithms. Comput. Methods Appl. Mech. Eng. **186**, 311–338 (2000)
6. Lampinen, J.: A constraint handling approach for the differential evolution algorithm. In: Proceedings of the IEEE Congress on Evolutionary Computation, vol. 2, pp. 1468–1473 (2002)
7. He, Q., Wang, L.: A hybrid particle swarm optimization with a feasibility-based rule for constrained optimization. Appl. Math. Comput. **186**, 1407–1422 (2007)
8. Hu, X., Eberhart, R.: Solving constrained nonlinear optimization problems with particle swarm optimization. In: Proceedings of the 6th World Multi-conference on Systemics, Cybernetics and Informatics (2002)
9. Koziel, S., Michalewicz, Z.: Evolutionary algorithms, homomorphous mapping, and constrained parameter optimization. Evol. Comput. **7**(1), 19–44 (1999)
10. Coello Coello, C.A., Becerra, R.L.: Efficient evolutionary optimization through the use of a cultural algorithm. Eng. Optim. **36**(2), 219–236 (2004)
11. Becerra, R.L., Coello Coello, C.A.: Cultured differential evolution for constrained optimization. Comput. Methods Appl. Mech. Eng. **195**, 4303–4322 (2006)
12. Chootinan, P., Chen, A.: Constraint handling in genetic algorithms using a gradient-based repair method. Comput. Oper. Res. **33**, 2263–2281 (2006)
13. Zahara, E., Hu, C.H.: Solving constrained optimization problems with hybrid particle swarm optimization. Eng. Optim. **40**(11), 1031–1049 (2008)
14. Dong, Y., Tang, J., Xu, B., Wang, D.: An application of swarm optimization to nonlinear programming. Comput. Math. Appl. **49**, 1655–1668 (2005)
15. Hedar, A.R., Fukushima, M.: Derivative-free simulated annealing method for constrained continuous global optimization. J. Global Optim. **35**, 521–549 (2006)
16. Coello Coello, C.A., Montes, E.M.: Constraint-handling in genetic algorithms through the use of dominance-based tournament selection. Adv. Eng. Inform. **16**, 193–203 (2002)
17. Deb, K.: GeneAS: a robust optimal design technique for mechanical component design. In: Dasgupta, D., Michalewicz, Z., (eds.) Evolutionary Algorithms in Engineering Applications, pp. 497–514. Springer, New York (1997)
18. Kannan, B.K., Kramer, S.N.: An augmented lagrange multiplier based method for mixed integer discrete continuous optimization and its applications to mechanical design. ASME J. Mech. Des. **116**, 405–411 (1994)
19. Ragsdell, K.M., Phillips, D.T.: Optimal design of a class of welded structures using geometric programming. ASME J. Eng. Ind. Ser. B **98**(3), 1021–1025 (1976)
20. Sandgren, E.: Nonlinear integer and discrete programming in mechanical design. In: Proceedings of the ASME Design Technology Conference, pp. 95–105 (1988)
21. Runarsson, T.P., Yao, X.: Stochastic ranking for constrained evolutionary optimization. IEEE Trans. Evol. Comput. **4**(3), 284–294 (2000)
22. Hamida, S.B., Schoenauer, M.: ASCHEA: new results using adaptive segregational constraint handling. In: Fogel, D.B., et al. (eds.) Proceedings of the IEEE Congress on Evolutionary Computation, pp. 884–889 (2002)

23. Montes, E.M., Coello Coello, C.A.: A simple multimembered evolution strategy to solve constrained optimization problems. Technical Report EVOCINV-04-2003, Evolutionary Computation Group at CINVESTAV, Sección de Computación, Departamento de Ingeniería Eléctrica, CINVESTAV-IPN, México D.F., México
24. Ray, T., Tai, K., Seow, K.C.: An evolutionary algorithm for constrained optimization. In: Proceedings of the Genetic and Evolutionary Computation Conference, pp. 771–777 (2000)
25. Arora, J.S.: Introduction to Optimum Design. Elsevier Academic Press, San Diego (2004)
26. He, Q., Wang, L.: An effective co-evolutionary particle swarm optimization for constrained engineering design problems. Eng. Appl. Artif. Intell. **20**, 89–99 (2006)
27. Siddall, J.N.: Analytical Design-Making in Engineering Design. Prentice-Hall, Englewood Cliffs (1972)

Chapter 4
Modified Cohort Intelligence for Solving Machine Learning Problems

Clustering is an important and popular technique in data mining. It partitions a set of objects in such a manner that objects in the same clusters are more similar to each another than objects in the different cluster according to certain predefined criteria. K-means is simple yet an efficient method used in data clustering. However, K-means has a tendency to converge to local optima and depends on initial value of cluster centers. In the past, many heuristic algorithms have been introduced to overcome this local optima problem. Nevertheless, these algorithms too suffer several short-comings. In this chapter, we present an efficient hybrid evolutionary data clustering algorithm referred as to K-MCI, whereby, we combine K-means with modified cohort intelligence. Our proposed algorithm is tested on several standard data sets from UCI Machine Learning Repository and its performance is compared with other well-known algorithms such as K-means, K-means++, cohort intelligence (CI), modified cohort intelligence (MCI), genetic algorithm (GA), simulated annealing (SA), tabu search (TS), ant colony optimization (ACO), honey bee mating optimization (HBMO) and particle swarm optimization (PSO). The simulation results are very promising in the terms of quality of solution and convergence speed of algorithm.

4.1 Introduction

Clustering is an unsupervised classification technique which partitions a set of objects in such a way that objects in the same clusters are more similar to one another than the objects in different clusters according to certain predefined criterion [1, 2]. The term unsupervised means that grouping is establish based on the intrinsic structure of the data, without any need to supply the process with training items.

Clustering has been applied across many applications, i.e., machine learning [3, 4], image processing [5–8], data mining [9, 10], pattern recognition [11, 12],

A.J. Kulkarni et al., *Cohort Intelligence: A Socio-inspired Optimization Method*,
Intelligent Systems Reference Library 114, DOI 10.1007/978-3-319-44254-9_4

bioinformatics [13–15], construction management [16], marketing [17, 18], document clustering [19], intrusion detection [19], healthcare [20, 21] and information retrieval [22, 23].

Clustering algorithms can generally be divided into two categories; hierarchical clustering and partitional clustering [24]. Hierarchical clustering groups objects into tree-like structure using bottom-up or top-down approaches. Our research however focuses on partition clustering, which decomposes the data set into a several disjoint clusters that are optimal in terms of some predefined criteria.

There many algorithms have been proposed in literature to solve the clustering problems. The K-means algorithm is the most popular and widely used algorithm in partitional clustering. Although, K-means is very fast and simple algorithm, it suffers two major drawbacks. Firstly, the performance of K-means algorithm is highly dependent on the initial values of cluster centers. Secondly, the objective function of the K-means is non-convex and it may contain many local minima. Therefore, in the process of minimizing the objective function, the solution might easily converge to a local minimum rather than a global minimum [25]. K-means++ algorithm was proposed by Arthur and Vassilvitskii [26], which introduces a cluster centers initialization procedure to tackle the initial centers sensitivity problem of a standard K-means. However, it too suffers from a premature convergence to a local optimum.

In order to alleviate the local minima problem, many heuristic clustering approaches have been proposed over the years. For instance, [27] proposed a simulated annealing approach for solving clustering problems. A tabu search method which combines new procedures called packing and releasing was employed to avoid local optima in clustering problems [28]. Genetic algorithm based clustering method was introduced by Maulik and Bandyopadhyay [29] to improve the global searching capability. Fathian et al. [30] proposed a honey-bee mating optimization approach for solving clustering problems. Shelokar et al. [31] proposed an ant colony optimization (ACO) for clustering problems. A particle swarm optimization based approach (PSO) for clustering was introduced by Chen and Ye [32] and Cura [33]. A hybrid technique for clustering called KNM-PSO, which combines the K-means, Nedler-Mead simplex and PSO was proposed by Kao et al. [34]. Zhang et al. [35] proposed an artificial bee colony approach for clustering. More recently, black hole (BH) optimization algorithm [36] was introduced to solve clustering problems. Although these heuristic algorithms address the flaws of K-means but they still suffer several drawbacks. For example, most of these heuristic algorithms are typically very slow to find optimum solution. Furthermore, these algorithms are computationally expensive for large problems.

Cohort intelligence (CI) is a novel optimization algorithm proposed recently by Kulkarni et al. [37]. This algorithm was inspired from natural and society tendency of cohort individuals/candidates of learning from one another. The learning refers to a cohort candidate's effort to self-supervise its behavior and further adapt to the behavior of other candidate which it tends to follow. This makes every candidate to improve/evolve its own and eventually the entire cohort behavior. CI was tested with several standard problems and compared with other optimization algorithms such as sequential quadratic programming (SQP), chaos-PSO (CPSO), robust hybrid PSO (RHPSO) and linearly decreasing weight PSO (LDWPSO). CI has been proven to be computationally comparable and even better performed in terms of quality of solution and computational efficiency when compared with these algorithms. These comparisons can be found in the seminal paper on CI [37]. However, for clustering problems, as the number of clusters and dimensionality of data increase, CI might converge slowly and trapped in local optima. Recently, many researchers have incorporated mutation operator into their algorithm to solve combinatorial optimizing problems. Several new variants of ACO algorithms have been proposed by introducing mutation to the traditional ACO algorithms and achieve much better performance [38, 39]. Stacey et al. [40] and Zhao et al. [39] also have integrated mutation into the standard PSO scheme, or modifications of it. In order to mitigate the short-comings of CI, we present a modified cohort intelligence (MCI) by incorporating mutation operator into CI to enlarge the searching range and avoid early convergence. Finally, to utilize the benefits of both K-means and MCI, we propose a new hybrid K-MCI algorithm for clustering. In this algorithm, K-means is applied to improve the candidates' behavior that generated by MCI at each iteration before going through the mutation process of MCI. The new proposed hybrid K-MCI is not only able to produce a better quality solutions but it also converges more quickly than other heuristic algorithms including CI and MCI. In summary, our contribution in this chapter is twofold:

1. Present a modified cohort intelligence (MCI).
2. Present a new hybrid K-MCI algorithm for data clustering.

4.2 The Clustering Problem and K-Means Algorithm

Let $R = [Y_1, Y_2, \ldots, Y_N]$, where $Y_i \in \Re^D$, be a set of N data objects to be clustered and $S = [X_1, X_2, \ldots, X_K]$ be a set of K clusters. In clustering, each data in set R will be allocated in one of the K clusters in such a way that it will minimize the objective function. The objective function, intra-cluster variance is defined as the sum of squared Euclidean distance between each object Y_i and the center of the cluster X_j which it belongs. This objective function is given by:

$$F(X,Y) = \sum_{i=1}^{N} Min\left\{\left\|Y_i - X_j\right\|^2\right\}, \quad j = 1, 2, \ldots, K \tag{4.1}$$

Also,

- $X_j \neq \emptyset, \forall j \in \{1, 2, \ldots, K\}$
- $X_i \cap X_j = \emptyset, \forall i \neq j$ and $\forall i, j \in \{1, 2, \ldots, K\}$
- $\cup_{j=1}^{K} X_j = R$

In partitional clustering, the main goal of K-means algorithm is to determine centers of K clusters. In this research, we assume that the number of clusters K is known prior to solving the clustering problem. The following are the main steps of K-means algorithm:

- Randomly choose K cluster centers of X_1, X_2, …, X_K from data set $R = [Y_1, Y_2, \ldots, Y_N]$ as the initial centers.
- Assign each object in set R to the closest centers.
- When all objects have been assigned, recalculate the positions of the K centers.
- Repeat Step 2 and 3 until a termination criterion is met (the maximum number of iterations reached or the means are fixed).

Arthur and Vassilvitskii [26] introduced a specific way of choosing the initial centers for the K-means algorithm. The procedure of the K-means++ algorithm is outlined below:

- Choose one center X_1, uniformly at random from R.
- For each data point Y_i, compute $D\ (Y_i)$, the distance between Y_i and the nearest center that has already been chosen.
- Take new center X_j, choosing $Y \in R$ with probability $\frac{D(Y)^2}{\sum_{Y \in R} D(Y)^2}$.
- Repeat Steps 2 and 3 until K centers have been chosen.
- Now that the initial centers have been chosen, proceed using standard K-means clustering.

4.3 Modified Cohort Intelligence

In this chapter, we present a modified cohort intelligence (MCI) to improve the accuracy and the convergence speed of CI. Premature convergence may arise when the cohort converges to a local optimum or the searching process of algorithm is very slow. Therefore, we introduced a mutation mechanism to CI in order to enlarge the searching range, expand the diversity of solutions and avoid early convergence.

Assume for ith iteration, a candidate in a particular cohort is represented by a set of K number of cluster centers, $S^c = [X_1^c, X_2^c, \ldots, X_j^c, \ldots, X_K^c]$, where $c = 1, 2, \ldots, C$ and X_j^c represents the cluster's center. For an example, Fig. 4.1 depicts a candidate solution of a problem with three clusters, $K = 3$ and all the data objects have four dimensions, $D = 4$. Thus, the candidate solution illustrated in Fig. 4.1 can be represented by $S^c = [x_1^c, x_2^c, \ldots, x_b^c]_{1 \times b}$, where $b = K \times D$. Then, each candidate S^c in that cohort will undergo mutation process to generate mutant candidate S_{mut}^c as following:

$$S_{mut}^c = S^{m1} + rand(.) * (S^{m2} - S^{m3}) \tag{4.2}$$

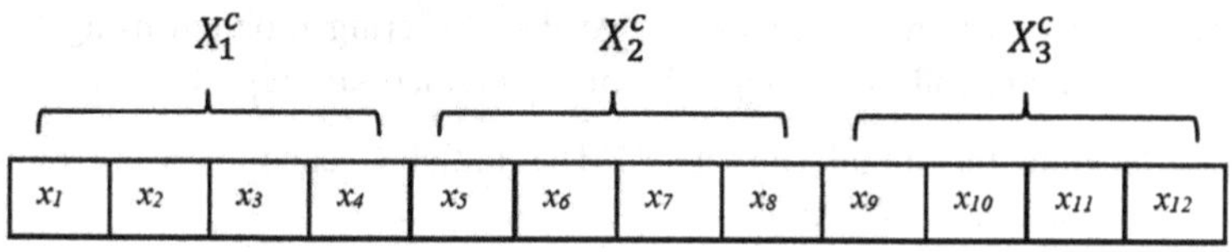

Fig. 4.1 Example of a candidate solution

Variables $m1$, $m2$ and $m3$ are three candidates which are selected randomly from C candidates in such a way that $m1 \neq m2 \neq m3 \neq c$.

$$S^c_{mut} = \left[x^c_{mut,1}, x^c_{mut,2}, \ldots, x^c_{mut,b}\right]_{1\times b} \tag{4.3}$$

The selected candidate would be:

$$S^c_{trial} = \left[x^c_{trial,1}, x^c_{trial,2}, \ldots, x^c_{trial,b}\right]_{1\times b} \tag{4.4}$$

$$x^c_{trial,z} = \begin{cases} x^c_{mut,z} & if \;\; rand(.) < \gamma \\ x^c_z \end{cases} \tag{4.5}$$

where $z = 1, 2, \ldots, b$, $rand(.)$ is a random number between 0 and 1, γ is a random number less than 1 and D is the dimensionality of data objects. Thus, the new features for candidate c in the ith iteration are selected based on its objective function:

$$S^c_{new} = \begin{cases} S^c & if \;\; f(S^c) \leq f(S^c_{trial}) \\ S^c_{trial} & otherwise \end{cases} \tag{4.6}$$

This mutation process is performed to other remaining candidates in cohort.

4.4 Hybrid K-MCI and Its Application for Clustering

In this chapter, we propose a novel algorithm referred to as the hybrid K-means modified cohort intelligence (K-MCI) for data clustering. In this algorithm, K-means is utilized to improve the candidates' behavior generated by MCI. After a series run of K-means, then each candidate will go through the mutation process as described in Sect. 4.3. The new proposed algorithm benefits from the advantages of both K-means and MCI. This combination allows the proposed algorithm to converge more quickly and achieve a more accurate solutions without getting trapped to a local optimum. The application of the hybrid K-MCI on the data clustering is presented in this section. In order to solve the clustering problem using the new proposed algorithm, following steps should be applied and repeated:

Step 1. Generate the initial candidates. The initial C candidates are randomly generated as described below:

$$Candidates = \begin{bmatrix} S^1 \\ S^2 \\ \vdots \\ S^c \\ \vdots \\ S^C \end{bmatrix} \tag{4.7}$$

$$S^c = [X_1^c, X_2^c, \ldots, X_K^C] \tag{4.8}$$

$$X_j^c = [x_1^c, x_2^c, \ldots, x_D^c] \tag{4.9}$$

where c = 1, 2, …, C, K is the number of clusters, j = 1, 2, …, K and D is the dimensionality of cluster center X_j^c.

Thus,

$$S^c = \left[x_1^c, x_2^c, \ldots, x_i^c, \ldots, x_b^c\right]_{1\times b}, \quad where\ b = K \times D \tag{4.10}$$

The sampling interval ψ_i is given by $x_i^{c,\min} \le x_i \le x_i \le x_i^{c,\max}$, where, $x_i^{c,\min}$ and $x_i^{c,\max}$ (each feature of center) are minimum and maximum value of each point belonging to the cluster X_j^c.

Step 2. Perform K-means algorithm for each candidate as described in Sect. 4.2.

Step 3. Perform mutation operation for each candidate as described in Sect. 4.3.

Step 4. The objective function $f(S^c)$ for each candidate is calculated.

Step 5. The probability of selecting the behavior $f^*(S^c)$ of every candidate is calculated.

Step 6. Every candidate generates a random number *rand* [0, 1] and by using the roulette wheel approach decides to follow corresponding behavior $f^*(S^{c[?]})$ and its features $S^{c[?]} = [x_1^{c[?]}, x_2^{c[?]}, \ldots, x_b^{c[?]}]$. For example, candidate c [1] may decide to follow behavior of candidate $f^*(S^{c[2]})$ and its features $S^{c[2]} = [x_1^{c[2]}, x_2^{c[2]}, \ldots, x_b^{c[2]}]$.

Step 7. Every candidate shrinks the sampling interval $\psi^{c[?]}$ for its every features $x_i^{c[?]}$ to its local neighborhood.

Step 8. Each candidate samples t qualities from within the updated sampling interval of its selected features,$x_i^{c[?]}$. Then, each candidate computes the objective function for these t behaviors and selects the best behavior $f^*(S^c)$ from this set. For instance with t = 15, candidate c [1] decides to follow the behavior of candidate $f^*(S^{c[2]})$ and its features $S^{c[2]} = [x_1^{c[2]}, x_2^{c[2]}, \ldots, x_b^{c[2]}]$. Then, candidate c [1] will sample 15 qualities from its updated sampling interval features of $x_i^{c[2]}$. Next, candidate c [1] will compute the objective function of its behaviors according, i.e. $F^{c[1]} = [f(S^{c[1]})^1, f(S^{c[1]})^2, \ldots, f(S^{c[1]})^{15}]$ and selects the best behavior $f^*(S^{c[1]})$ from within this set.

Step 9. Accept any of the C behaviors from current set of behaviors in the cohort as the final objective function value $f^*(S)$ and its features $S^c = [x_1^c, x_2^c, \ldots, x_b^c]$ and stop if either of the two criteria listed below is valid or else continue to Step 2:

1. If maximum number of iterations exceeded.
2. If cohort saturates to the same behavior by satisfying the conditions convergence condition.

The flow chart of the hybrid K-MCI is illustrated in Fig. 4.2.

4.5 Experiment Results

Six real data sets are used to validate our proposed algorithm. Each data set from UCI Machine Learning Repository has a different number of clusters, data objects and features as described below [41]:

Iris data set (N = 150, D = 4, K = 3): which consists of three different species of Iris flowers: *Iris setosa, Iris versicolour* and *Iris virginica.* For each species, 50 samples with four features (sepal length, sepal width, petal length, and petal width) were collected.

Wine data set (N = 178, D = 13, K = 3): This data set are the results of a chemical analysis of wines grown in the same region in Italy but derived from three different cultivators: class 1 (59 instances), class 2 (71 instances), and class 3 (48 instances). The analysis determined the quantities of 13 features found in each of the three types of wines. These 13 features are alcohol, malic acid, ash, alkalinity of ash, magnesium, total phenols, flavanoids, nonflavanoid phenols, proanthocyanins, color intensity, hue, OD280/OD315 of diluted wines, and proline.

Glass data set (N = 214, D = 9, K = 6): which consists of six different types of glass: building windows float processed (70 objects), building windows non-float processed (76 objects), vehicle windows float processed (17 objects), containers (13 objects), tableware (9 objects), and headlamps (29 objects). Each type of glass has nine features, which are refractive index, sodium, magnesium, aluminum, silicon, potassium, calcium, barium, and iron.

Breast Cancer Wisconsin data set (N = 683, D = 9, K = 2): This data set contains 683 objects. There are two categories: malignant (444 objects) and benign

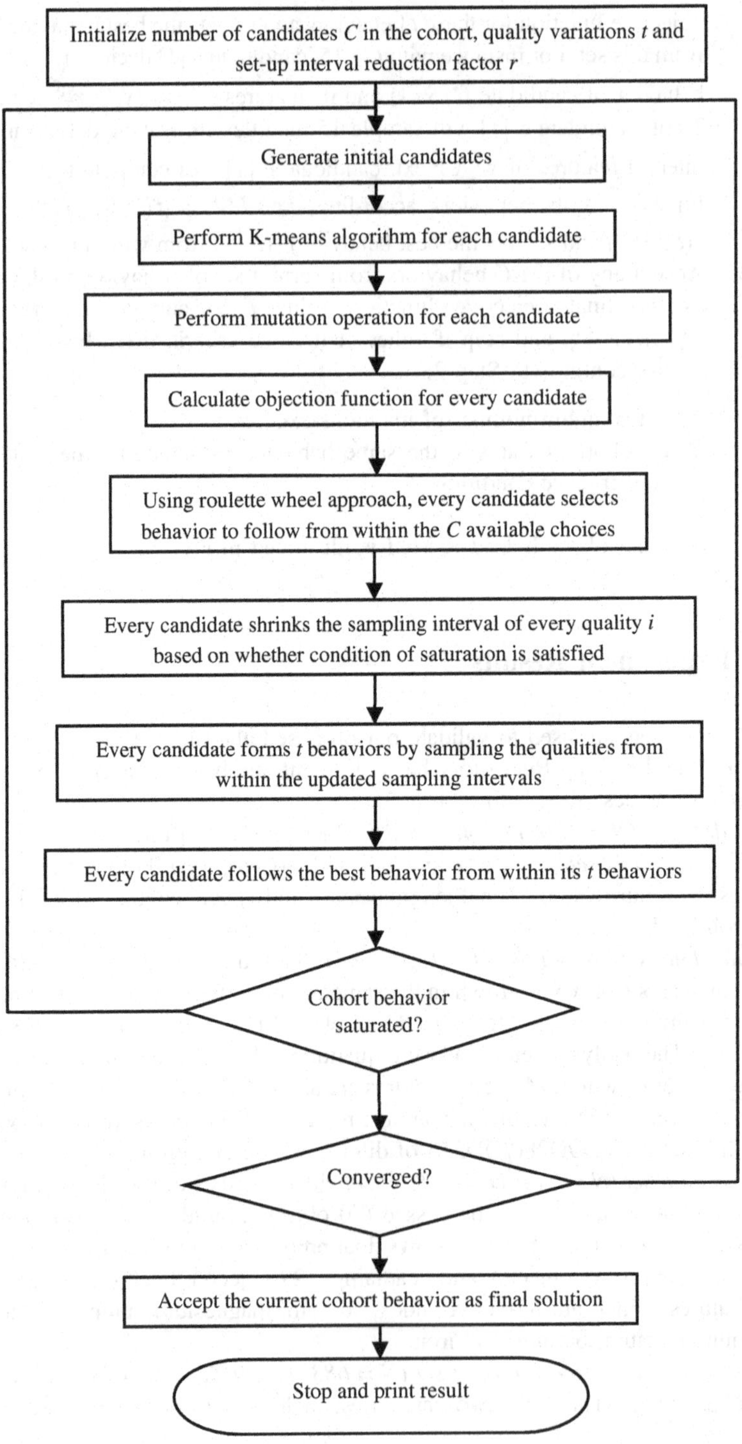

Fig. 4.2 The flow chart of the hybrid K-MCI

(239 objects). Each type of class consists of nine features, which includes clump thickness, cell size uniformity, cell shape uniformity, marginal adhesion, single epithelial cell size, bare nuclei, bland chromatin, normal nucleoli and mitoses.

Vowel data set ($N = 871$, $D = 3$, $K = 6$): which consist of 871 Indian Telugu vowels sounds. There are six-overlapping vowel classes: δ (72 instances), a (89 instances), I (172 instances), u (151 instances), e (207 instances) and o (180 instances). Each class has three input features corresponding to the first, second, and third vowel frequencies.

Contraceptive Method Choice data set ($N = 1473$, $D = 9$, $K = 3$): This data set is a subset of the 1987 National Indonesia Contraceptive Prevalence Survey. The samples are married women who either were not pregnant or did not know if they were at the time of interview. The problem is to predict the choice of current contraceptive method (no use has 629 objects, long-term methods have 334 objects, and short-term methods have 510 objects) of a woman based on her demographic and socioeconomic characteristics.

The performance of our proposed algorithm on these selected data set is compared with several typical stochastic algorithms such as the CI, MCI, ACO [31, 34], SA [27, 42], GA [29], TS [42], HBMO [43], K-means and K-means++. We have utilized two criteria to evaluate the performance of these algorithms: (i) the intra-cluster distances and (ii) the number of fitness function evaluation (NFE). For the first criteria, numerically smaller the value of the intra-cluster distances indicates higher the quality of the clustering is. As for the second criteria, the smaller NFE value indicates the high convergence speed of the considered algorithm.

The required parameters for the implementation of hybrid K-MCI, MCI and CI for clustering are shown in Table 4.1. The algorithms are implemented with Matlab 8.0 on a Windows platform using Intel Core i7-3770, 3.4 GHz and 8 GB RAM computer. Table 4.2 shows the summary of the intra-cluster distances obtained by the clustering algorithms on the selected data sets. The results are best, average, worst and the standard deviation of solutions over 20 independent runs. The NFE criteria in Table 4.2 indicates convergence speed of the respective algorithms. NFE is the number of times that the clustering algorithm has calculated the objective function to reach the optimal solution.

The simulations results given in Table 4.2, shows that our proposed method performs much better than other methods for all test data sets. Our proposed method is able to achieve the best optimal value with a smaller standard deviation compared

Table 4.1 Parameters of hybrid K-MCI, MCI and CI for data clustering

Data	CI			MCI			K-MCI		
	t	v	r	t	v	r	t	v	r
Iris	5	15	0.95	5	15	0.95	5	15	0.92
Wine	5	15	0.95	5	15	0.95	5	15	0.7
Cancer	5	15	0.95	5	15	0.95	5	15	0.95
Vowel	5	15	0.99	5	15	0.99	5	15	0.98
CMC	5	15	0.99	5	15	0.99	5	15	0.99
Glass	5	15	0.99	5	15	0.99	5	15	0.98

Table 4.2 Simulation results for clustering algorithms

Dataset	Criteria	K-means	K-means++	GA	SA	TS	ACO	HBMO	PSO	CI	MCI	K-MCI
Iris	Best	97.3259	97.3259	113.9865	97.4573	97.3659	97.1007	96.7520	96.8942	96.6557	96.6554	96.6554
	Average	106.5766	98.5817	125.1970	99.9570	97.8680	97.1715	96.9531	97.2328	96.6561	96.6554	96.6554
	Worst	123.9695	122.2789	139.7782	102.0100	98.5694	97.8084	97.7576	97.8973	96.6570	96.6554	96.6554
	S.D.	12.938	5.578	14.563	2.018	0.53	0.367	0.531	0.347	0.0002	0	0
	NFE	80	71	38,128	5314	20,201	10,998	11,214	4953	7250	4500	3500
Wine	Best	16,555.68	16,555.68	16,530.53	16,473.48	16,666.22	16,530.53	16,357.28	16,345.96	16,298.01	16,295.16	16,292.44
	Average	17,251.35	16,816.55	16,530.53	17,521.09	16,785.45	16,530.53	16,357.28	16,417.47	16,300.98	16,296.51	16,292.7
	Worst	18,294.85	18,294.85	16,530.53	18,083.25	16,837.53	16,530.53	16,357.28	16,562.31	16,305.06	16,297.98	16,292.88
	S.D.	874.148	637.14	0	753.084	52.073	0	0	85.497	2.118	0.907	0.13
	NFE	285	261	33,551	17,264	22,716	15,473	7238	16,532	17,500	16,500	6250
Cancer	Best	2988.43	2986.96	2999.32	2993.45	2982.84	2970.49	2989.94	2973.5	2964.64	2964.4	2964.38
	Average	2988.99	2987.99	3249.46	3239.17	3251.37	3046.06	3112.42	3050.04	2964.78	2964.41	2964.38
	Worst	2999.19	2988.43	3427.43	3421.95	3434.16	3242.01	3210.78	3318.88	2964.96	2964.43	2964.38
	S.D.	2.469	0.689	229.734	230.192	232.217	90.5	103.471	110.801	0.094	0.007	0
	NFE	120	112	20,221	17,387	18,981	15,983	19,982	16,290	7500	7000	5000
CMC	Best	5703.2	5703.2	5705.63	5849.03	5885.06	5701.92	5699.26	5700.98	5695.33	5694.28	5693.73
	Mean	5704.57	5704.19	5756.59	5893.48	5993.59	5819.13	5713.98	5820.96	5696.01	5694.58	5693.75
	Worst	5705.37	5705.37	5812.64	5966.94	5999.8	5912.43	5725.35	5923.24	5696.89	5694.89	5693.8
	S.D.	1.033	0.955	50.369	50.867	40.845	45.634	12.69	46.959	0.482	0.198	0.014
	NFE	187	163	29,483	26,829	28,945	20,436	19,496	21,456	30,000	28,000	15,000

(continued)

Table 4.2 (continued)

Dataset	Criteria	K-means	K-means++	GA	SA	TS	ACO	HBMO	PSO	CI	MCI	K-MCI
Glass	Best	215.73	215.36	278.37	275.16	279.87	269.72	245.73	270.57	219.37	213.03	212.34
	Mean	218.7	217.56	282.32	282.19	283.79	273.46	247.71	275.71	223.31	214.08	212.57
	Worst	227.35	223.71	286.77	287.18	286.47	280.08	249.54	283.52	225.48	215.62	212.8
	S.D.	2.456	2.455	4.138	4.238	4.19	3.584	2.438	4.55	1.766	0.923	0.135
	NFE	533	510	199,892	199,438	199,574	196,581	195,439	198,765	55,000	50,000	25,000
Vowel	Best	149,398.66	149,394.56	149,513.73	149,370.47	149,468.26	149,395.6	149,201.63	148,976.01	149,139.86	148,985.35	148,967.24
	Mean	151,987.98	151,445.29	159,153.49	161,566.28	162,108.53	159,458.14	161,431.04	151,999.82	149,528.56	149,039.86	148,987.55
	Worst	162,455.69	161,845.54	165,991.65	165,986.42	165,996.42	165,939.82	165,804.67	158,121.18	150,468.36	149,102.38	149,048.58
	S.D.	3425.25	3119.751	3105.544	2847.085	2846.235	3485.381	2746.041	2881.346	495.059	43.735	36.086
	NFE	146	129	10,548	9423	9528	8046	8436	9635	15,000	13,500	7500

to other methods. In short, the results highlighted the precision and robustness of the proposed K-MCI as compared to other algorithms including CI and MCI. For Iris data set, K-MCI and MCI algorithm are able to converge to global optimum of 96.5554 for each run, while the best solutions for CI, K-Means, K-means++, GA, SA, TS, ACO, HBMO and PSO are 96.6557, 97.3259, 97.3259, 113.9865, 97.4573, 97.3659, 97.1007, 96.752 and 96.8942. The standard deviation for K-MCI is zero, which is much less than other methods. K-MCI is also able to achieve the best global result and has a better average and worst result for the Wine data set compared to other methods. As for CMC data set, K-MCI has the best solution of 5693.73, while the best solutions for CI, MCI, K-Means, K-means++, GA, SA, TS, ACO, HBMO and PSO are 5695.33, 5694.28, 5703.20, 5703.20, 5705.63, 5849.03, 5885.06, 5701.92, 5699.26 and 5700.98. Furthermore, KMCI has a much smaller standard deviation than the other methods for CMC data set. For vowel data set, our proposed method also manages to achieve best, average, worst solution and standard deviation of 148,967.24, 148,987.55, 149,048.58 and 36.086. These obtained values are much smaller than other methods.

We notice the effect of applying mutation operator to CI by comparing the results between MCI and CI from Table 4.2. For instance, MCI has achieved a best, average, worst solutions of 16,295.16, 16,296.51 and 16,297.98 with a standard deviation of 0.907 for Wine data set while CI has obtained best, average, worst solutions of 16,298.01, 16,300.98 and 16,305.60 with a standard deviation of 2.118. Thus, by applying mutation operator, MCI is able to produce a better quality solution compared to the original CI.

The simulation results from Table 4.2 for K-MCI, MCI and CI points out the advantages of hybridizing K-means into MCI. The best global solution of K-MCI, MCI and CI for the Wine data set are 16,292.44, 16,295.16 and 16,298.01. These results prove that K-MCI will provide a higher clustering quality than the standalone MCI and CI. Besides improving the clustering quality, the combination of K-means with MCI, will further enhance the convergence characteristics. CI and MCI need 17,500 and 16,500 function evaluations respectively to obtain the best solution for Wine data set. On the other hand, K-MCI only takes 6250 function evaluations to achieve the best optimal solution for the same data set. Hence, K-MCI converges to optimal solution very quickly. Although standalone K-means

Table 4.3 The achieved best centers on Cancer data set

Dataset	Center 1	Center 2
Cancer	7.11701	2.88942
	6.64106	1.12774
	6.62548	1.20072
	5.61469	1.16404
	5.24061	1.99334
	8.10094	1.12116
	6.07818	2.00537
	6.02147	1.10133
	2.32582	1.03162

Table 4.4 Achieved best centers on the glass and vowel data set

Dataset	Center 1	Center 2	Center 3	Center 4	Center 5	Center 6
Glass	1.52434	1.51956	1.51362	1.52132	1.51933	1.51567
	12.03344	13.25068	13.15690	13.74692	13.08412	14.65825
	0.01215	0.45229	0.65548	3.51952	3.52765	0.06326
	1.12869	1.53305	3.13123	1.01524	1.36555	2.21016
	71.98256	73.01401	70.50411	71.89517	72.85826	73.25324
	0.19252	0.38472	5.33024	0.21094	0.57913	0.02744
	14.34306	11.15803	6.73773	9.44764	8.36271	8.68548
	0.23039	0.00433	0.67322	0.03588	0.00837	1.02698
	0.15156	0.06599	0.01490	0.04680	0.06182	0.00382
Vowel	506.98650	623.71854	407.89515	439.24323	357.26154	375.45357
	1839.66652	1309.59677	1018.05210	987.68488	2291.44000	2149.40364
	2556.20000	2333.45721	2317.82688	2665.47618	2977.39697	2678.44208

Table 4.5 The archived best centers on the Iris, Wine and CMC data set

Dataset	Center 1	Center 2	Center 3
Iris	5.01213	5.93432	6.73334
	3.40309	2.79781	3.06785
	1.47163	4.41787	5.63008
	0.23540	1.41727	2.10679
Wine	13.81262	12.74160	12.50086
	1.83004	2.51921	2.48843
	2.42432	2.41113	2.43785
	17.01717	19.57418	21.43603
	105.41208	98.98807	92.55049
	2.93966	1.97496	2.02977
	3.21965	1.26308	1.54943
	0.34183	0.37480	0.32085
	1.87181	1.46902	1.38624
	5.75329	5.73752	4.38814
	1.05368	1.00197	0.94045
	2.89757	2.38197	2.43190
	1136.97230	687.01356	463.86513
CMC	43.64742	24.41296	33.50648
	2.99091	3.03823	3.13272
	3.44673	3.51059	3.55176
	4.59136	1.79036	3.65914
	0.80254	0.92502	0.79533
	0.76971	0.78935	0.69725
	1.82586	2.29463	2.10130
	3.42522	2.97378	3.28562
	0.10127	0.03692	0.06151
	1.67635	2.00149	2.11479

and K-means++ algorithms converge much faster than other algorithms including K-MCI, they have a tendency to prematurely converge to a local optimum. For instance, K-means++ algorithm only needs 261 function evaluations to obtain the best solution for Wine data set but these solution results are suboptimal.

In summary, the simulation results from Table 4.2 validates that our proposed method is able to attain a better global solution with a smaller standard deviation and fewer numbers of function evaluations for clustering. Finally, we have included Tables 4.3, 4.4 and 4.5 to illustrate the best centers found by K-MCI in the test data.

4.6 Conclusion

CI is a new emerging optimization method, which has a great potential to solve many optimization problems including for data clustering. However, CI may converge slowly and prematurely converge to local optima when the dimensionality of data and number of cluster centers increase. With the purpose of assuaging these drawbacks, we proposed modified CI (MCI) by implementing mutation operator into CI. It outperforms CI in terms of both quality of solutions and the convergence speed. Finally in this chapter, we proposed a novel hybrid K-MCI algorithm for data clustering. This new algorithm tries to exploit the merits of the two algorithms simultaneously, where the K-means is utilized to improve the candidates of MCI at each iteration before these candidates are given back again to MCI for optimization. This combination of K-means and MCI allows our proposed algorithm to convergence more quickly and prevents it from falling to local optima. We tested our proposed method using the standard data sets from UCI Machine Learning Repository and compared our results with six state-of-art clustering methods. The experimental results indicate that our proposed algorithm can produce a higher quality clusters with a smaller standard deviation on the selected data set compare to other clustering methods. Moreover, the convergence speed to global optima of the proposed algorithm is better than other heuristic algorithms. In other words, our proposed method can be considered as an efficient and reliable method to find the optimal solution for clustering problems.

There are a number of future research directions can be considered to improve and extend this research. The computational performance is governed by parameters such as the sampling interval reduction factor r. Thus, a self-adaptive scheme can be introduced to fine tune the sampling interval reduction. In this research, we assumed the number of clusters are known a prior when solving the clustering problems. Therefore, we can further modify our algorithm to perform automatic clustering without any prior knowledge of number of clusters. We may combine MCI with other heuristic algorithms to solve clustering problems, which can be seen as another research direction. Finally, our proposed algorithm may be applied to solve other practically important problems such as image segmentation [44], traveling salesman problem [45], process planning and scheduling [46] and load dispatch of power system [47].

References

1. Jain, A.K., Murty, M.N., Flynn, P.J.: Data clustering: a review. ACM Comput. Surv. **31**, 264323 (1999)
2. Kaufman, L., Rousseeuw, P.: Finding Groups in Data: An Introduction to Cluster Analysis (Wiley Series in Probability and Statistics). Wiley-Interscience (2005)
3. Fan, S., Chen, L., Lee, W.-J.: Machine learning based switching model for electricity load forecasting. Energy Convers. Manag. **49**, 1331–1344 (2008)
4. Anaya, A.R., Boticario, J.G.: Application of machine learning techniques to analyse student interactions and improve the collaboration process. Expert Syst. Appl. **38**, 1171–1181 (2011)
5. Das, S., Konar, A.: Automatic image pixel clustering with an improved differential evolution. Appl. Soft Comput. **9**, 226236 (2009)
6. Siang Tan, K., Mat Isa, N.A.: Color image segmentation using histogram thresholding fuzzy c-means hybrid approach. Pattern Recogn. **44**, 1–15 (2011)
7. Portela, N.M., Cavalcanti, G.D., Ren, T.I.: Semi-supervised clustering for MR brain image segmentation. Expert Syst. Appl. **41**, 1492–1497 (2014)
8. Zhao, F., Fan, J., Liu, H.: Optimal-selection-based suppressed fuzzy c-means clustering algorithm with self-tuning non local spatial information for image segmentation. Expert Syst. Appl. **41**, 4083–4093 (2014)
9. Ci, S., Guizani, M., Sharif, H.: Adaptive clustering in wireless sensor networks by mining sensor energy data. Netw. Coverage Routing Schemes Wirel. Sens. Netw. **30**, 2968–2975 (2007)
10. Carmona, C., Ramrez-Gallego, S., Torres, F., Bernal, E., del Jesus, M., Garca, S.: Web usage mining to improve the design of an ecommerce website: Orolivesur.com. Expert Syst. Appl. **39**, 11243–11249 (2012)
11. Yuan, T., Kuo, W.: Spatial defect pattern recognition on semiconductor wafers using model-based clustering and bayesian inference. Eur. J. Oper. Res. **190**, 228–240 (2008)
12. Bassiou, N., Kotropoulos, C.: Long distance bigram models applied to word clustering. Pattern Recogn. **44**, 145158 (2011)
13. Bhattacharya, A., De, R.K.: Average correlation clustering algorithm (ACCA) for grouping of co-regulated genes with similar pattern of variation in their expression values. J. Biomed. Inform. **43**, 560–568 (2010)
14. Macintyre, G., Bailey, J., Gustafsson, D., Haviv, I., Kowalczyk, A.: Using gene ontology annotations in exploratory microarray clustering to understand cancer etiology. Pattern Recogn. Lett. **31**, 2138–2146 (2010)
15. Zheng, B., Yoon, S.W., Lam, S.S.: Breast cancer diagnosis based on feature extraction using a hybrid of k-means and support vector machine algorithms. Expert Syst. Appl. **41**, 1476–1482 (2014)
16. Cheng, Y.-M., Leu, S.-S.: Constraint-based clustering and its applications in construction management. Expert Syst. Appl. **36**, 5761–5767 (2009)
17. Kuo, R., An, Y., Wang, H., Chung, W.: Integration of selforganizing feature maps neural network and genetic k-means algorithm for market segmentation. Expert Syst. Appl. **30**, 313–324 (2006)
18. Kim, K.-J., Ahn, H.: A recommender system using GA k-means clustering in an online shopping market. Expert Syst. Appl. **34**, 1200–1209 (2008)
19. Jun, S., Park, S.-S., Jang, D.-S.: Document clustering method using dimension reduction and support vector clustering to overcome sparseness. Expert Syst. Appl. **41**, 3204–3212 (2014)
20. Gunes, S., Polat, K., Yosunkaya, S.: Efficient sleep stage recognition system based on EEG signal using k-means clustering based feature weighting. Expert Syst. Appl. **37**, 7922–7928 (2010)
21. Hung, Y.-S., Chen, K.-L.B., Yang, C.-T., Deng, G.-F.: Web usage mining for analysing elder self-care behavior patterns. Expert Syst. Appl. **40**, 775–783 (2013)
22. Chan, C.-C.H.: Intelligent spider for information retrieval to support mining-based price prediction for online auctioning. Expert Syst. Appl. **34**, 347–356 (2008)

23. Dhanapal, R.: An intelligent information retrieval agent. Know. Based Syst. **21**, 466–470 (2008)
24. Han, J.: Data Mining: Concepts and Techniques. Morgan Kaufmann Publishers Inc. (2005)
25. Selim, S.Z., Ismail, M.A.: K-means-type algorithms: a generalized convergence theorem and characterization of local optimality. IEEE Trans. Pattern Anal. Mach. Intell. **6**, 81–87 (1984)
26. Arthur, D., Vassilvitskii, S.: K-means++: the advantages of careful seeding. Proceedings of the Eighteenth Annual ACM-SIAM Symposium on Discrete Algorithms SODA '07, pp. 1027–1035. Society for Industrial and Applied Mathematics, Philadelphia, PA, USA (2007)
27. Selim, S.Z., Alsultan, K.: A simulated annealing algorithm for the clustering problem. Pattern Recogn. **24**, 1003–1008 (1991)
28. Sung, C., Jin, H.: A tabu-search-based heuristic for clustering. Pattern Recogn. **33**, 849–858 (2000)
29. Maulik, U., Bandyopadhyay, S.: Genetic algorithm-based clustering technique. Pattern Recogn. **33**, 1455–1465 (2000)
30. Fathian, M., Amiri, B., Maroosi, A.: Application of honey-bee mating optimization algorithm on clustering. Appl. Math. Comput. **190**, 1502–1513 (2007)
31. Shelokar, P.S., Jayaraman, V.K., Kulkarni, B.D.: An ant colony approach for clustering. AnalyticaChimicaActa **509**, 187–195 (2004)
32. Chen, C.-Y., Ye, F.: Particle swarm optimization algorithm and its application to clustering analysis. In: IEEE International Conference on Networking, Sensing and Control, vol. 2, pp. 789–794 (2004)
33. Cura, T.: A particle swarm optimization approach to clustering. Expert Syst. Appl. **39**, 1582–1588 (2012)
34. Kao, Y.-T., Zahara, E., Kao, I.-W.: A hybridized approach to data clustering. Expert Syst. Appl. **34**, 1754–1762 (2008)
35. Zhang, C., Ouyang, D., Ning, J.: An artificial bee colony approach for clustering. Expert Syst. Appl. **37**, 4761–4767 (2010)
36. Hatamlou, A.: Black hole: a new heuristic optimization approach for data clustering. Inf. Sci. **222**, 175–184 (2013)
37. Kulkarni, A.J., Durugkar, I.P., Kumar, M.: Cohort intelligence: a self supervised learning behavior. In: 2013 IEEE International Conference on Systems, Man, and Cybernetics (SMC), pp. 1396–1400 (2013)
38. Lee, Z.-J., Su, S.-F., Chuang, C.-C., Liu, K.-H.: Genetic algorithm with ant colony optimization (ga-aco) for multiple sequence alignment. Appl. Soft Comput. **8**, 55–78 (2008)
39. Zhao, N., Wu, Z., Zhao, Y., Quan, T.: Ant colony optimization algorithm with mutation mechanism and its applications. Expert Syst. Appl. **37**, 4805–4810 (2010)
40. Stacey, A., Jancic, M., Grundy, I.: Particle swarm optimization with mutation. In: The 2003 Congress on Evolutionary Computation, 2003 (CEC '03), vol. 2, pp. 1425–1430 (2003)
41. Bache, K., Lichman, M.: UCI Machine Learning Repository. University of California, Irvine, School of Information and Computer Sciences (2013)
42. Niknam, T., Amiri, B.: An efficient hybrid approach based on PSO, ACO and k-means for cluster analysis. Appl. Soft Comput. **10**, 183–197 (2010)
43. Fathian, M., Amiri, B.: A honeybee-mating approach for cluster analysis. Int. J. Adv. Manuf. Technol. **38**, 809–821 (2008)
44. Bhandari, A.K., Singh, V.K., Kumar, A., Singh, G.K.: Cuckoo search algorithm and wind driven optimization based study of satellite image segmentation for multilevel thresholding using kapurs entropy. Expert Syst. Appl. **41**, 3538–3560 (2014)
45. Albayrak, M., Allahverdi, N.: Development a new mutation operator to solve the traveling salesman problem by aid of genetic algorithms. Expert Syst. Appl. **38**, 1313–1320 (2011)
46. Seker, A., Erol, S., Botsali, R.: A neuro-fuzzy model for a new hybrid integrated process planning and scheduling system. Expert Syst. Appl. **40**, 5341–5351 (2013)
47. Zhisheng, Z.: Quantum-behaved particle swarm optimization algorithm for economic load dispatch of power system. Expert Syst. Appl. **37**, 1800–1803 (2010)

Chapter 5
Solution to 0–1 Knapsack Problem Using Cohort Intelligence Algorithm

The previous chapters discussed the algorithm Cohort Intelligence (CI) and its applicability solving several unconstrained and constrained problems. In addition CI was also applied for solving several clustering problems. This validated the learning and self supervising behavior of the cohort. This chapter further tests the ability of CI by solving an NP-hard combinatorial problem such as Knapsack Problem (KP). Several cases of the 0–1 KP are solved. The effect of various parameters on the solution quality has been discussed. The advantages and limitations of the CI methodology are also discussed.

5.1 Knapsack Problem Using CI Method

The Knapsack Problem (KP) can be divided into two categories, Single-constraint KPs and Multiple-constraint KPs. The single-constraint KPs include the Subset-sum, 0–1 Knapsack, Bounded Knapsack, change-making, and Multiple-choice Knapsack. On the other hand, the multiple-constraint KPs include 0–1 Multiple Knapsack, 0–1 Multidimensional Knapsack, generalized assignment, and Bin Packing with a wide range of applications, such as cargo loading, cutting stock problems, resource allocation in computer systems, and economics [1, 2]. The special case of single constraints is generally known as the KP or the Uni-dimensional KP [3]. Another variant of the KP referred to as Multichoice Multidimensional KP (MMKP) is used to represent an optimally graceful Quality of Service (QoS) degradation model where the QoS of a single session multimedia service is gracefully degraded to conform to changes in resource availability [4]. Khan [5] used the MMKP to represent a utility model (UM) which is a mathematical model for a multi-session adaptive multimedia system. The MMKP also appears in the nursing personnel scheduling problem [6], which is defined as the identification of a staffing pattern that specifies the number of nursing personnel of a certain skill to be scheduled and satisfies the total nursing personnel capacity and

A.J. Kulkarni et al., *Cohort Intelligence: A Socio-inspired Optimization Method*,
Intelligent Systems Reference Library 114, DOI 10.1007/978-3-319-44254-9_5

other relevant constraints. Practical applications of 0–1 Knapsack include finding an optimal investment plan [7], as well as theoretical applications such as a sub-problem when solving generalized assignment problem, which is heavily used when solving vehicle routing problems, efficient packing of cargo containers by considering the weight and volume capacity utilization, etc. [8]. Apart from these applications, KPs are being used for resource allocation problems dealing with the World Wide Web [9]. In this chapter various cases of the 0–1 KP [10–12] were solved using CI. In all the problems, the implemented CI methodology produced robust results with reasonable computational cost.

The problem is described as follows [10–14]: given a set of N objects, each object i, $i = 1, \ldots, N$ is associated with an integer profit v_i and an integer weight w_i. Fill the knapsack with a subset of the objects such that the total profit $f(\mathbf{v})$ is maximized and the total weight $f(\mathbf{w})$ does not exceed a given capacity W. The mathematical formulation is as follows:

$$\text{Maximize}\, f(\mathbf{v}) = \sum_{i=1}^{N} v_i x_i$$

$$\text{Subject to}\, f(\mathbf{w}) \leq W$$

where

$$f(\mathbf{w}) = \sum_{i=1}^{N} w_i x_i, x_i \in \{0, 1\}, \quad 1 \leq i \leq N \tag{5.1}$$

5.1.1 *Illustration of CI Solving 0–1 KP*

In the context of CI algorithm (discussed in Sect. 5.1), the objects $i, i = 1, \ldots, N$ are considered as characteristics/attributes/qualities which decide the overall profit $f(\mathbf{v})$ and the associated overall weight $f(\mathbf{w})$ of the knapsack. The procedure begins with the initialization of the number of cohort candidates C, and the number of variations t. In the cohort of C candidates, initially every candidate $c(c = 1, \ldots, C)$ randomly selects few objects, and the associated profits $\mathbf{F}^C = \{f(\mathbf{v}^1), \ldots, f(\mathbf{v}^c), \ldots, f(\mathbf{v}^C)\}$ and weights $\mathbf{F}^{CW} = \{f(\mathbf{w}^1), \ldots, f(\mathbf{w}^c), \ldots, f(\mathbf{w}^C)\}$ are calculated. The further CI algorithm steps are discussed below.

Step 1. The probability $p^c(c = 1, \ldots, C)$ of selecting a profit $f(\mathbf{v}^c)$, $(c = 1, \ldots, C)$, is calculated as $p^c = p_1^c + p_2^c$ where

$$p_1^c = \frac{f(\mathbf{v}^c)}{\sum_{c=1}^{C} f(\mathbf{v}^c)} \tag{5.2}$$

and

$$p_2^c = \begin{cases} \frac{f(\mathbf{w}^c)}{W} & f(\mathbf{w}^c) \leq W \\ 3 - \frac{2f(\mathbf{w}^c)}{W} & f(\mathbf{w}^c) > W \end{cases} \tag{5.3}$$

A probability distribution specially devised to bias the solution towards feasibility is represented in Fig. 5.1. The Probability p_2^c increases linearly as the total weight of the knapsack increases, and reaches its peak value at the maximum capacity W. Upon any further increase in weight the probability rapidly decreases. Thus, the probability is highest around maximum capacity and decreases on either side of it with decrease beyond W with twice the slope.

Step 2. Based on roulette wheel selection approach every candidate $c(c = 1, \ldots, C)$ selects a candidate with associated profit $f\left(\mathbf{v}^{c[?]}\right)$ and modifies its own solution by incorporating some objects from that candidate. The superscript [?] indicates that the behavior is selected by candidate c and not known in advance. The modification approach is inspired from the feasibility-based rules discussed in [15–17]. The modifications are categorized as follows:

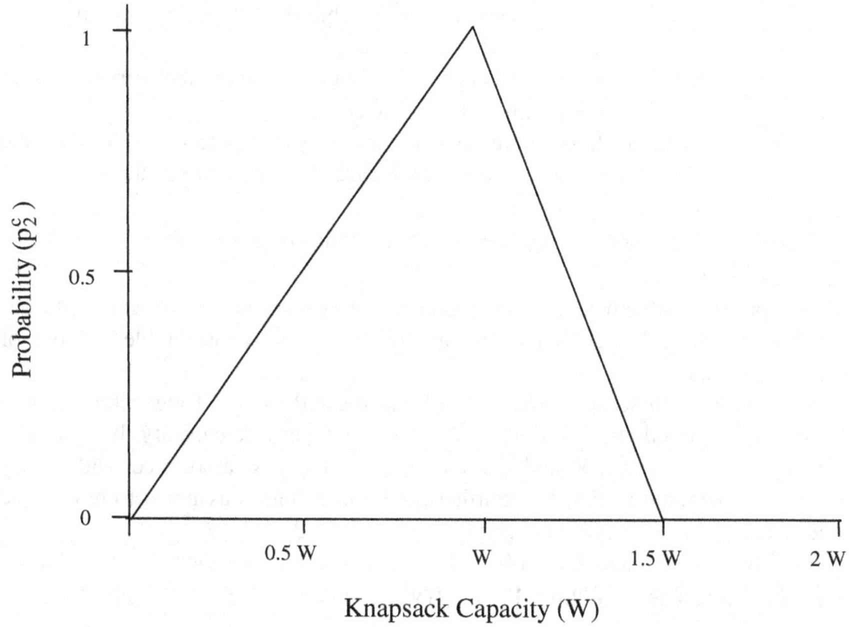

Fig. 5.1 Probability distribution for p_2^c

1. If the solution of candidate c $(c = 1,\ldots,C)$ is feasible i.e. it satisfies the weight constraint given by Eq. 5.1 then, it randomly chooses one of the following modifications:
   1.1. Adds a randomly chosen object from the candidate being followed, such that the object has not been included in the present candidate c and the weight constraint given by Eq. 5.1 is still satisfied.
   1.2. Replaces a randomly chosen object with another randomly chosen one from the candidate being followed, such that Eq. 5.1 is satisfied.
2. If the candidate c $(c = 1,\ldots,C)$ is infeasible then, it randomly chooses one of the following modifications:
   2.1. Removes a randomly chosen object from within its knapsack.
   2.2. Replaces a randomly chosen object with another randomly chosen one from the candidate c being followed, such that the total weight $f(\mathbf{w}^c)$ of the candidate c decreases.
   Every candidate performs the above procedure t times. This makes every candidate c available with associated profits $\mathbf{F}^{c,t} = \left\{f(\mathbf{v}^c)^1,\ldots,f(\mathbf{v}^c)^j,\ldots,f(\mathbf{v}^c)^t\right\}$, $(c = 1,\ldots,C)$. Furthermore, every candidate selects the best profit $f^*(\mathbf{v})$ among them. The best variation is selected based on the following conditions:
      2.2.1. If the variations are feasible then the variation with maximum profit is selected.
      2.2.2. If the variations are infeasible then the variation with minimum weight is selected.
      2.2.3. If there are both infeasible and feasible variations then the feasible variation with maximum profit is selected.

This makes the cohort available with C updated profits $\mathbf{F}^C = \{f^*(\mathbf{v}^1),\ldots, f^*(\mathbf{v}^c),\ldots,f^*(\mathbf{v}^C)\}$.

This process continues until saturation (convergence) i.e., every candidate has the same profit and it does not change for successive considerable number of learning attempts.

The above discussed procedure of solving the KP using CI algorithm is illustrated here with number of objects $N = 4$ and knapsack capacity $W = 8$. The weights w_i, $i = 1,\ldots,N$ and profits v_i, $i = 1,\ldots,N$ associated with every object are illustrated in Fig. 5.2. Furthermore, the cohort is assumed to have three candidates, i.e. $C = 3$ and number of variations $t = 3$.

Initially every candidate $c(c = 1,\ldots,C)$ randomly selects few objects, and the associated profits $\mathbf{F}^C = \{f(\mathbf{v}^1),\ldots,f(\mathbf{v}^c),\ldots,f(\mathbf{v}^C)\}$ and weights

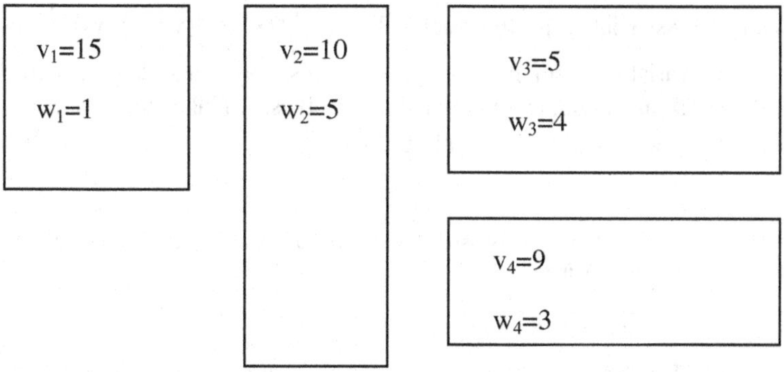

Fig. 5.2 Illustrative 0–1 KP example with $N = 4, W = 8$

$\mathbf{F}^{C,W} = \{f(\mathbf{w}^1), \ldots, f(\mathbf{w}^c), \ldots, f(\mathbf{w}^C)\}$ are calculated. The further CI algorithm steps are discussed below:

(1) The probability p^c associated with each candidate $c(c = 1, \ldots, 3)$ is calculated using Eqs. 5.2 and 5.3. The calculated probability values are presented in Fig. 5.3.
(2) Using roulette wheel selection approach, assume that candidate 1 decides to follow candidate 3. As presented in Fig. 5.4, $t = 3$ variations are formed along

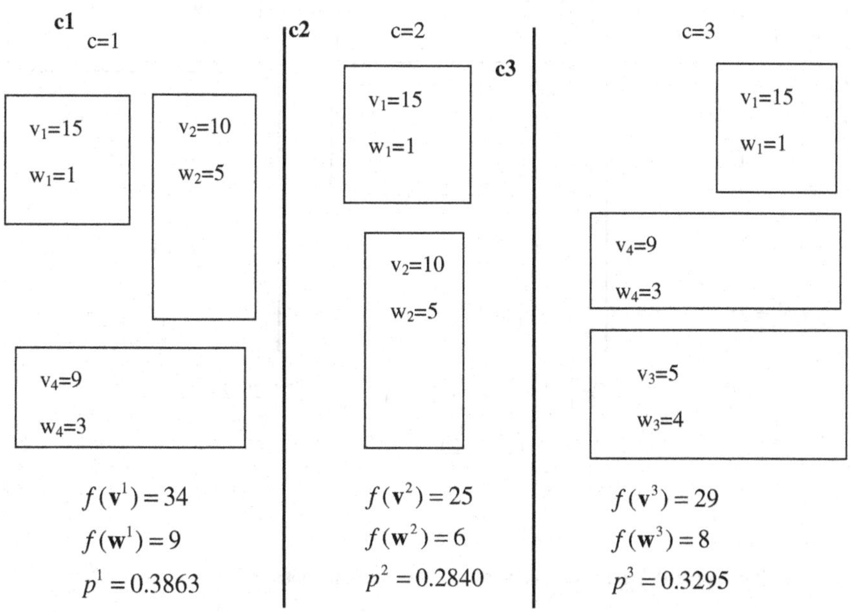

Fig. 5.3 Illustrative 0–1 KP example with $C = 3$

with the associated profits vector $\mathbf{F}^{1,3} = \left\{ f(\mathbf{v}^1)^1, f(\mathbf{v}^1)^2, f(\mathbf{v}^1)^3 \right\}$ and the selected variation with profit $f^*(\mathbf{v}^1)$. In this way, candidates 2 and 3 also follow certain candidate and update themselves. It makes the cohort available with 3 updated candidates with profits $\mathbf{F}^3 = \{ f^*(\mathbf{v}^1), f^*(\mathbf{v}^2), f^*(\mathbf{v}^3) \}$. This

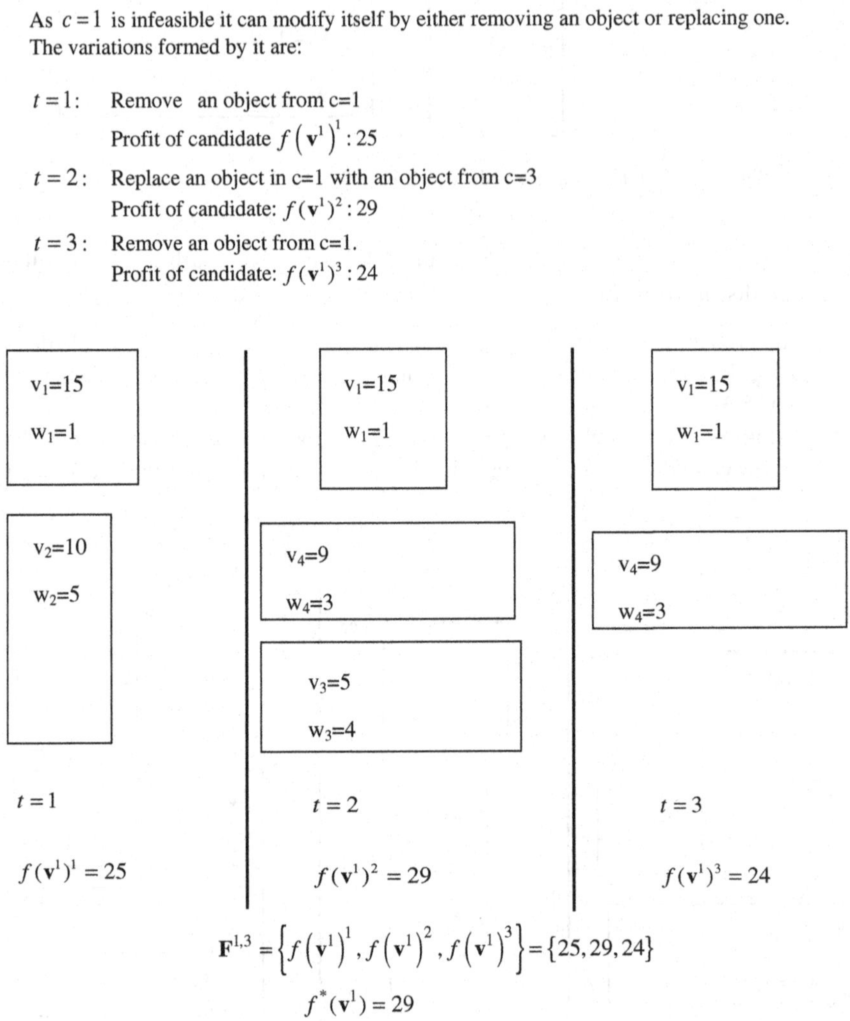

Fig. 5.4 Illustrative 0–1 KP example with $t = 3$ (variations obtained)

process continues until saturation (convergence), i.e. every candidate finds the solution and does not change for successive considerable number of learning attempts.

5.2 Results and Discussion

The CI algorithm discussed in Sect. 5.1 was coded in Matlab 7.11 (R2010b) and the simulations were run on a Windows platform using an Intel Core i5 CPU, 2.27 GHz processor speed and 3 GB memory capacity, and further validated using twenty distinct test cases of the 0–1 Knapsack Problems. The standard test cases $f_1 - f_{10}$ [10–12] are presented in Table 5.1. The cases $f_{11} - f_{20}$ were generated using a random number generator. In these tests, knapsack capacity is calculated using the formula [11, 12]: $W = \frac{3}{4}\sum_{i=1}^{N} w_i$ where w_i is a random weight of item i and N is the number of items. Different values of N were used, varying from 30 to 75. These test cases are presented at the end of this chapter.

Recently, the instances $f_1 - f_{10}$ were solved using Harmony Search (HS) [10, 13], Improved Harmony Search (IHS) [10, 14], Novel Global Harmony Search (NGHS) [10–12], Quantum Inspired Cuckoo Search Algorithm QICSA [12], and Quantum Inspired Harmony Search Algorithm (QIHSA) [11]. The HS is based on natural musical performance processes and has been applied to a variety of engineering problems; however, it exhibits poor convergence rate [10]. IHS employs a parameter updating method for generating new solution vectors that enhances accuracy and convergence rate of HS algorithm. The convergence rate is further improved in NGHS which is inspired from the swarm intelligence and employs a dynamic updating strategy and probabilistic mutation approach; however, the performance degenerates significantly when applied for solving constrained problems. All these algorithms lack a method to satisfy constraints and hence, can result in an infeasible solution when solving constrained optimization problems. Zou et al. [10] used a penalty function method along with NGHS in order to handle the weight constraint in 0–1 KP. QICSA integrates the quantum computing principles such as qubit representation, measure operation and quantum mutation, in the Cuckoo Search algorithm. It is different from other evolutionary algorithms in that it offers a large exploration of the search space through intensification and diversification [12]. QIHSA combines the features of HS algorithm and quantum computing. The probabilistic nature of the quantum measure offers a good diversity to the harmony search algorithm, while the interference operation helps to intensify the search around the best solutions [11]. While hybridization between quantum inspired computing and nature inspired algorithms significantly improve the performance over the original nature inspired algorithms, their performance depends largely on the initial solutions, which are selected randomly. Also, when dealing with constrained optimization problems they require the use of a repair operator.

Table 5.1 The dimension and parameters of ten test problems

f	Number of objects (N)	Parameters ($W, \mathbf{w}, \mathbf{v}$)
f_1	10	$W = 269$, $\mathbf{w} = \{95, 4, 60, 32, 23, 72, 80, 62, 65, 46\}$, $\mathbf{v} = \{55, 10, 47, 5, 4, 50, 8, 61, 85, 87\}$
f_2	20	$W = 878$, $\mathbf{w} = \{92, 4, 43, 83, 84, 68, 92, 82, 6, 44, 32, 18, 56, 83, 25, 96, 70, 48, 14, 58\}$, $\mathbf{v} = \{44, 46, 90, 72, 91, 40, 75, 35, 8, 54, 78, 40, 7, 15, 61, 17, 75, 29, 75, 63\}$
f_3	4	$W = 20$, $\mathbf{w} = \{6, 5, 9, 7\}$, $\mathbf{v} = \{9, 11, 13, 15\}$
f_4	4	$W = 11$, $\mathbf{w} = \{2, 4, 6, 7\}$, $\mathbf{v} = \{6, 10, 12, 13\}$
f_5	15	$W = 375$, $\mathbf{w} = \{56.358531, 80.874050, 47.987304, 89.596240, 74.660482, 85.894345, 51.353496, 1.498459, 36.445204, 16.589862, 44.569231, 0.466933, 37.788018, 57.118442, 60.716575\}$, $\mathbf{v} = \{0.125126, 19.330424, 58.500931, 35.029145, 82.284005, 17.410810, 71.050142, 30.399487, 9.140294, 14.731285, 98.852504, 11.908322, 0.891140, 53.166295, 60.176397\}$
f_6	10	$W = 60$, $\mathbf{w} = \{30, 25, 20, 18, 17, 11, 5, 2, 1, 1\}$, $\mathbf{v} = \{20, 18, 17, 15, 15, 10, 5, 3, 1, 1\}$
f_7	7	$W = 50$, $\mathbf{w} = \{31, 10, 20, 19, 4, 3, 6\}$, $\mathbf{v} = \{70, 20, 39, 37, 7, 5, 10\}$
f_8	23	$W = 10{,}000$, $\mathbf{w} = \{983, 982, 981, 980, 979, 978, 488, 976, 972, 486, 486, 972, 972, 485, 485, 969, 966, 483, 964, 963, 961, 958, 959\}$, $\mathbf{v} = \{981, 980, 979, 978, 977, 976, 487, 974, 970, 485, 485, 970, 970, 484, 484, 976, 974, 482, 962, 961, 959, 958, 857\}$

(continued)

Table 5.1 (continued)

f	Number of objects (N)	Parameters ($W, \mathbf{w}, \mathbf{v}$)
f_9	5	$W = 80,\ \mathbf{w} = \{15, 20, 17, 8, 31\}$, $\mathbf{v} = \{33, 24, 36, 37, 12\}$
f_{10}	20	$W = 879,\ \mathbf{w} = \{84, 83, 43, 4, 44, 6, 82, 92, 25, 83, 56, 18, 58, 14, 48, 70, 96, 32, 68, 92\}$, $\mathbf{v} = \{91, 72, 90, 46, 55, 8, 35, 75, 61, 15, 77, 40, 63, 75, 29, 75, 17, 78, 40, 44\}$

The approach of CI handles constraints using a probability distribution p_2^c (refer to Fig. 5.1) which forces the candidates to follow the behaviour/solution with constraints satisfied as well as closer to the ones with constraint values closer to the boundary. Moreover, a well-established feasibility-based rule [15–17] was also incorporated which assists candidates select the variations with better objective and

Table 5.2 Summary of solutions of KPs solved using CI

Problem	Number of objects, Knapsack capacity (N, W)	Solution $(f^*(\mathbf{v}), f^*(\mathbf{w}))$			Standard deviation	Average function evaluations (FE)	Average time (s)	Parameters (C, t)
		Best	Mean	Worst				
f_1	10, 269	295, 269	267.46, 262.722	260, 250	0.0	5410	0.4489	5, 10
f_2	20, 878	1024, 871	1020.55, 852.84	1009, 827	0.0	5446	1.5909	5, 10
f_3	4, 20	35, 18	34.55, 17.867	28, 16	0.0	5136	0.2687	5, 10
f_4	4, 11	23, 11	22.06, 10.33	16, 6	0.64	5193	0.2492	5, 10
f_5	15, 375	481.0694, 354.9608	449.986, 361.692	412.6988, 372.9118	10.68	5590	0.6609	5, 10
f_6	10, 60	51, 56	50.733, 56.733	49, 54	0.66	5573	0.4465	5, 10
f_7	7, 50	105, 50	86.6, 44.8	79, 42	2.99	5696	0.3749	5, 10
f_8	23, 10,000	9759, 9760	9753.33, 9756.33	9710, 9711	11.5	6486	1.1959	5, 10
f_9	5, 80	130, 60	124.6, 61.4	106, 74	2.89	5110	0.3048	5, 10
f_{10}	20, 879	1025, 871	997.7, 558.3	892, 805	18.6	5426	1.535	5, 10
f_{11}	30, 577	1437, 566	1418, 571.5	1398, 563	11.79	6817	3.4635	5, 10
f_{12}	35, 655	1689, 650	1686.5, 650.833	1679, 654	3.8188	5375	5.2288	5, 10
f_{13}	40, 819	1816, 817	1807.5, 817.66	1791, 819	9.604	7833	7.3429	5, 10
f_{14}	45, 907	2020, 903	2017, 902.5	2007, 901	4.749	7433	8.1510	5, 10
f_{15}	50, 882	2440, 873	2436.166, 870.33	2421, 865	6.841	7766	10.5690	5, 10
f_{16}	55, 1050	2643, 1049	2605, 1047.8	2581, 1049	22.018	9720	14.3445	5, 10
f_{17}	60, 1006	2917, 1002	2915, 1001.833	2905, 1001	4.472	9017	17.0894	5, 10
f_{18}	65, 1319	2814, 1319	2773.66, 1316.33	2716, 1317	18.273	10,283	20.9486	5, 10
f_{19}	70, 1426	3221, 1426	3216, 1423.166	3211, 1419	4.3589	10,333	26.4846	5, 10
f_{20}	75, 1433	3614, 1432	3603.8, 1431.8	3591, 1429	8.035	12,720	34.0072	5, 10

Table 5.3 Comparison of results obtained using CI with other established methods

Problem	Number of objects (N)	Method	Optimum solution $f^*(\mathbf{v})$
f_1	10	HS [10] IHS [10] NGHS [10, 11] QICSA [11] QIHSA [12] CI	295 295 295 295 295 295
f_2	20	HS [10] IHS [10] NGHS [10, 11] QICSA [11] QIHSA [12] CI	1024 1024 1024 1024 1024 1024
f_3	4	HS [10] IHS [10] NGHS [10, 11] QICSA [11] QIHSA [12] CI	35 35 35 35 35 35
f_4	4	HS [10] IHS [10] NGHS [10, 11] QICSA [11] QIHSA [12] CI	23 23 23 23 23 23
f_5	15	HS [10] IHS [10] NGHS [10, 11] QICSA [11] QIHSA [12] CI	481.0694 481.0694 481.0694 481.0694 481.0694 481.0694
f_6	10	HS [10] IHS [10] NGHS [10, 11] QICSA [11] QIHSA [12] CI	50 50 52 52 52 51
f_7	7	HS [10] IHS [10] NGHS [10, 11] QICSA [11] QIHSA [12] CI	107 107 107 107 107 105
f_8	23	HS [10] IHS [10] NGHS [10, 11] QICSA [11] QIHSA [12] CI	9767 9767 9767 9767 9767 9759

(continued)

Table 5.3 (continued)

Problem	Number of objects (N)	Method	Optimum solution $f^*(\mathbf{v})$
f_9	5	HS [10]	130
		IHS [10]	130
		NGHS [10, 11]	130
		QICSA [11]	130
		QIHSA [12]	130
		CI	130
f_{10}	20	HS [10]	1025
		IHS [10]	1025
		NGHS [10, 11]	1025
		QICSA [11]	1025
		QIHSA [12]	1025
		CI	1025

constraint satisfaction (refer to Sect. 5.1). The summary of the CI results including the best, mean and worst solutions with the associated average CPU time, average number of function evaluations, standard deviation are listed in Table 5.2. In addition, the CI parameters such as number of candidates C and number of variations t are also listed. As presented in Table 5.3, it can be seen that the solution was comparable for all problems and in most of the cases the optimum solution was obtained. In addition, according to Table 5.2, it is clear that the solution was obtained in reasonable computational cost (time and FE). The results have also been verified with Branch and Bound method, and according to Tables 5.3, 5.4 and Fig. 5.5 it is clear that the performance of CI and Branch and Bound are quite comparable. The CI saturation/convergence plot for one of the problems, $f_{10}(N = 20)$ is presented in Fig. 5.6 which illustrates the self adaptive learning behavior of every candidate in the cohort. Initially, the distinct behavior of every individual candidate in the cohort can be easily distinguished. As every candidate adopts the qualities of other candidates to improve its own solution, the cohort saturates to a certain improved solution. It is noted that the standard deviation was quite narrow with smaller sized problems; however, increased as the problem size increased. In addition, computational cost, i.e. time and function evaluations also increased with increase in the problem size. However, it was observed that in few runs of CI the candidates converged at suboptimal solutions. Similar to the perturbation approach implemented by Tavares et al. [17], in order to make the candidates jump out of possible local minima, every candidate $c(c = 1, \ldots, C)$ randomly selects a candidate to follow without considering its effect on the solution. This approach instantaneously made the solution worse, however, it was found to be helpful to pull the candidates' solution out of local minima and reach an improved solution. This approach was much simpler as opposed to the perturbation approach discussed by Tavares et al. [17] where several parameters were required to be tuned based on the preliminary trials.

Table 5.4 Comparison of results obtained using CI with B&B

Problem	Number of objects (N)	Optimum solution ($f^*(\mathbf{v})$) CI B&B	Time (s)
f_1	10	295 295	0.4489 0.12
f_2	20	1024 1024	1.5909 0.04
f_3	4	35 35	0.2687 0.03
f_4	4	23 23	0.2492 0.03
f_5	15	481.0694 481.0690	0.6609 0.18
f_6	10	51 52	0.4465 0.14
f_7	7	105 107	0.3749 0.04
f_8	23	9759 9767	1.1959 0.18
f_9	5	130 130	0.3048 0.03
f_{10}	20	1025 1025	1.535 0.45
f_{11}	30	1437 1437	3.4635 0.156001
f_{12}	35	1689 1689	5.2288 0.0624004
f_{13}	40	1816 1821	7.3429 0.0156001
f_{14}	45	2020 2033	8.1510 0.0312002
f_{15}	50	2440 2440	10.5690 0.0312002
f_{16}	55	2643 2440	14.3445 0.0312002
f_{17}	60	2917 2917	17.0894 0.0312002
f_{18}	65	2814 2818	20.9486 0.0624004
f_{19}	70	3221 3223	26.4846 0.0780005
f_{20}	75	3614 3614	34.0072 0.0312002

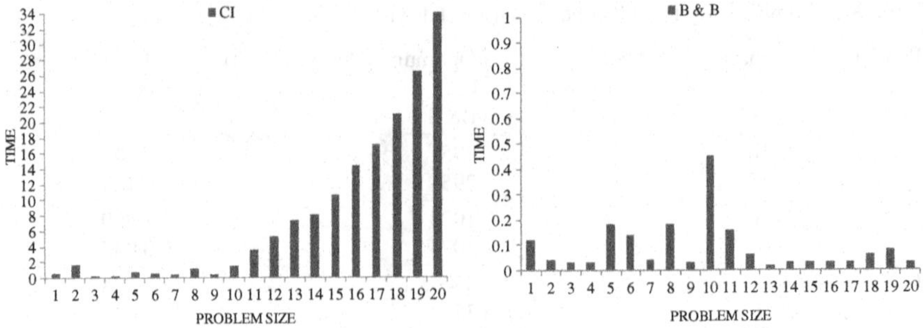

(a) CI (b) B&B

Fig. 5.5 Effect of problem size (N) on computational time

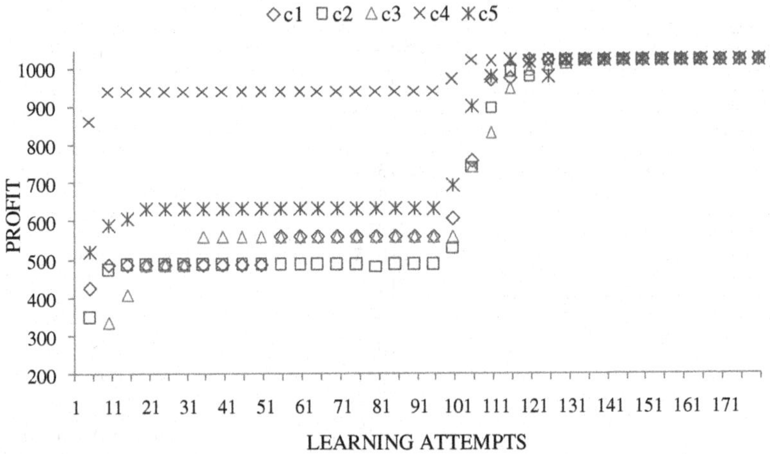

Fig. 5.6 Learning attempts versus the objective function $f(v)^*$ for each candidate

The effect of CI parameters viz. the number of candidates C and the number of variations in behaviour t was analyzed using the final values of profit $f(\mathbf{v})^*$, the total number of function evaluations and the computational time, for each problem. For every pair of number of candidates C and the number of variations in behavior t every KP test case was solved 20 times. For all the problems, the computational cost, i.e. the number of function evaluations and computational time was observed to be increasing with increasing number of candidates C, as well as number of variations in behaviour t. This was because, with increase in the number of candidates C and variations t, the number of behavior choices i.e. number of function evaluations also increased. The average values of profit $f(\mathbf{v})^*$, the total number of function evaluations and the computational time for different values of number of candidates C and variations t are shown for problem f_1 in Figs. 5.7, 5.8 and 5.9, respectively. Another important observation was that as the problem size i.e.

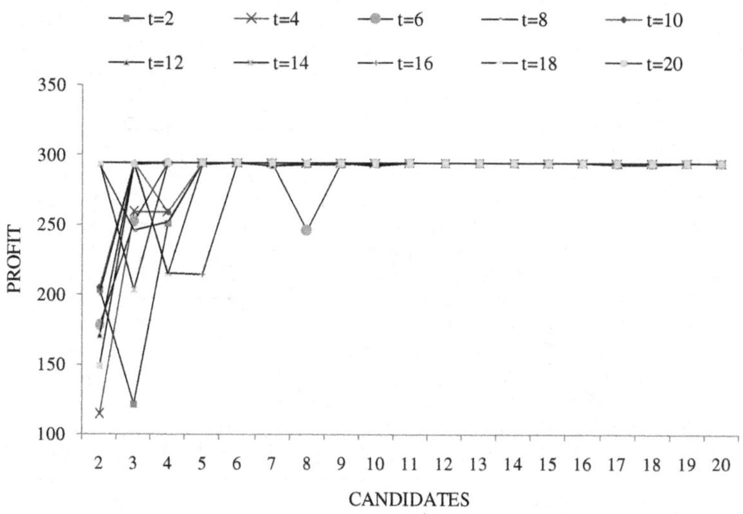

Fig. 5.7 Effect of number of candidates (C) on the profit $f(\mathbf{v})^*$ for different values of variations (t)

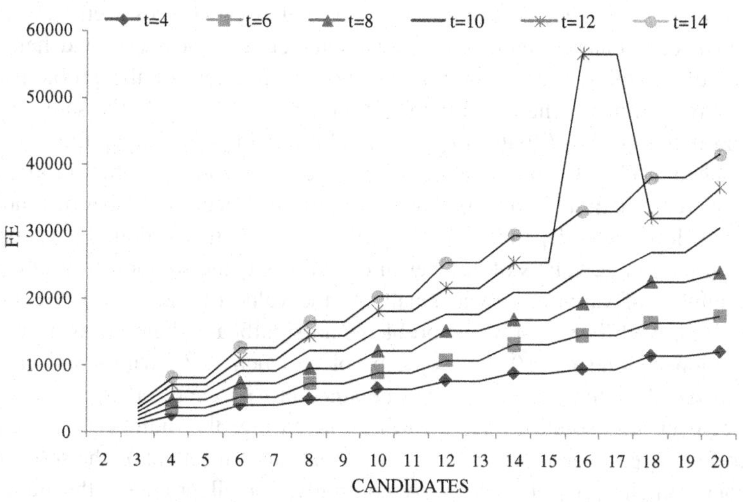

Fig. 5.8 Effect of number of candidates (C) on the function evaluations (FE) for different values of variations (t)

N increased, the computational cost also increased (refer to Table 5.2). Therefore, problems with larger number of objects took a longer time and more number of function evaluations to converge. Furthermore, with fewer number of candidates C, the solution, i.e. total profit $f(\mathbf{v})^*$ was suboptimal. As the value of number of candidates C was increased the solution quality improved up to a certain point after

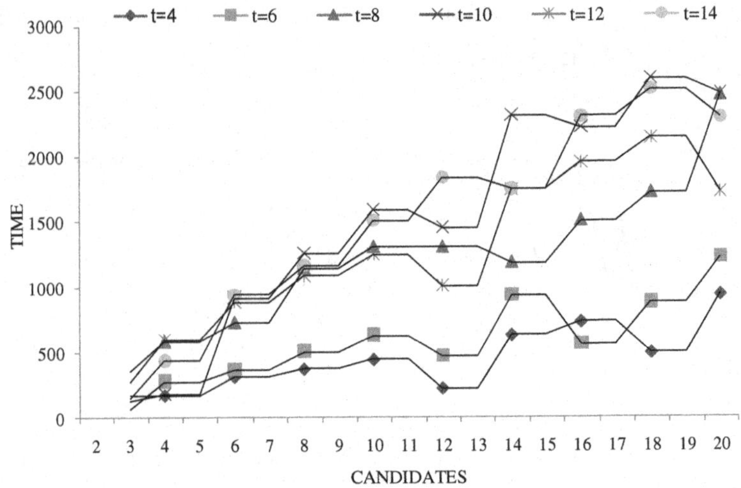

Fig. 5.9 Effect of number of candidates (C) on the time for different values of variations (t)

which there was no significant change (refer to Fig. 5.7). This was because for small values of number of candidates C the behavior choices were few and as number of candidates C increased the behavior choices increased and hence, the chances of selecting a better solution increased. In most of the problems there wasn't any significant change in the solution beyond $C = 5$. At the same time for some problems such as f_1 with small values of problem size N the optimum solution was reached at $C = 3$ and no significant change was observed in the solution upon further increase in number of candidates C. Thus, the effect of number of candidates C on the solution was dependent on the problem size N. In addition, it was observed that even with large values of number of candidates C the solution was suboptimal if the number of variations t was small. As the value of t increased, the solution quality improved. For most of the problems no significant change was observed in the solution beyond $t = 10$. For some problems such as f_1, with small values of problem size N optimum solution was obtained at $t = 4$ and no significant change was seen in the solution upon further increasing the number of variations t. Therefore, even in case of the number of variations t, its effect on the solution was dependent on the problem size N. Accordingly, for all problems the number of candidates C and number of variations t were chosen to be 5 and 10, respectively.

5.3 Conclusions and Future Directions

For the first time emerging CI algorithm has been applied for solving a combinatorial NP-hard problem such as 0–1 KP, with number of objects varying from 4 to 75. In all the problems the implemented CI methodology produced satisfactory results with reasonable computational cost. Furthermore, according to the solution comparison of CI with other contemporary methods it could be seen that the CI solution is comparable and for some problems even better than the other methods. The CI methodology was therefore validated and the self supervising nature of the cohort candidates was successfully demonstrated along with their ability to learn and improve qualities which further improved their individual behavior. In addition, in order to avoid saturation of cohort at suboptimal solution and further make the cohort saturate to the optimum solution, a generic approach such as accepting random behavior was incorporated. The effect of the important parameters such as number of candidates C and the associated variations t on the computational time, function evaluations and the solution was analysed. This could be a useful reference in dealing with future problems using CI.

It was observed that the computational time and function evaluations of the CI algorithm increased considerably with the problem size, in the future a self-adaptive scheme could be developed for these parameters such as number of candidates C and number of variations t. This may make CI algorithm computationally more efficient and improve the rate of convergence. In addition, authors also intend to further modify the CI algorithm to solve complex NP-hard bilevel programming problems from supply chain optimization domain [18]. Also, it is quite important to tune up the learning rate of CI candidates so as to apply to dynamic control systems [19]. The ability of CI in clustering [20–22] and classification domain in association with the cross-border transportation system and goods consolidation is currently underway.

5.4 Test Cases

f_{11}. N = 30, W = 577

$$\begin{aligned} \mathrm{w} = \{&46, 17, 35, 1, 26, 17, 17, 48, 38, 17, 32, 21, 29, 48, 31,\\ &8, 42, 37, 6, 9, 15, 22, 27, 14, 42, 40, 14, 31, 6, 34\}\\ \mathrm{v} = \{&57, 64, 50, 6, 52, 6, 85, 60, 70, 65, 63, 96, 18, 48, 85,\\ &50, 77, 18, 70, 92, 17, 43, 5, 23, 67, 88, 35, 3, 91, 48\} \end{aligned}$$

f_{12}. N = 35, W = 655

$$\begin{aligned} \mathrm{w} = \{&7, 4, 36, 47, 6, 33, 8, 35, 32, 3, 40, 50, 22, 18, 3, 12, 30, 31,\\ &13, 33, 4, 48, 5, 17, 33, 26, 27, 19, 39, 15, 33, 47, 17, 41, 40\}\\ \mathrm{v} = \{&35, 67, 30, 69, 40, 40, 21, 73, 82, 93, 52, 20, 61, 20, 42, 86, 43,\\ &93, 38, 70, 59, 11, 42, 93, 6, 39, 25, 23, 36, 93, 51, 81, 36, 46, 96\} \end{aligned}$$

f_{13}. N = 40, W = 819

$$w = \{28, 23, 35, 38, 20, 29, 11, 48, 26, 14, 12, 48, 35, 36, 33, 39, 30, 26, 44, 20, 13, 15, 46, 36, 43, 19, 32, 2, 47, 24, 26, 39, 17, 32, 17, 16, 33, 22, 6, 12\}$$
$$v = \{13, 16, 42, 69, 66, 68, 1, 13, 77, 85, 75, 95, 92, 23, 51, 79, 53, 62, 56, 74, 7, 50, 23, 34, 56, 75, 42, 51, 13, 22, 30, 45, 25, 27, 90, 59, 94, 62, 26, 11\}$$

f_{14}. N = 45, W = 907

$$w = \{18, 12, 38, 12, 23, 13, 18, 46, 1, 7, 20, 43, 11, 47, 49, 19, 50, 7, 39, 29, 32, 25, 12, 8, 32, 41, 34, 24, 48, 30, 12, 35, 17, 38, 50, 14, 47, 35, 5, 13, 47, 24, 45, 39, 1\}$$
$$v = \{98, 70, 66, 33, 2, 58, 4, 27, 20, 45, 77, 63, 32, 30, 8, 18, 73, 9, 92, 43, 8, 58, 84, 35, 78, 71, 60, 38, 40, 43, 43, 22, 50, 4, 57, 5, 88, 87, 34, 98, 96, 99, 16, 1, 25\}$$

f_{15}. N = 50, W = 882

$$w = \{15, 40, 22, 28, 50, 35, 49, 5, 45, 3, 7, 32, 19, 16, 40, 16, 31, 24, 15, 42, 29, 4, 14, 9, 29, 11, 25, 37, 48, 39, 5, 47, 49, 31, 48, 17, 46, 1, 25, 8, 16, 9, 30, 33, 18, 3, 3, 3, 4, 1\}$$
$$v = \{78, 69, 87, 59, 63, 12, 22, 4, 45, 33, 29, 50, 19, 94, 95, 60, 1, 91, 69, 8, 100, , 84, 100, 32, 81, 47, 59, 48, 56, 18, 59, 16, 45, 54, 4798, 75, 20, 4, 19, 58, 63, 37, 64, 90, 26, 29, 13, 53, 83\}$$

f_{16}. N = 55, W = 1050

$$w = \{27, 15, 46, 5, 40, 9, 36, 12, 11, 11, 49, 20, 32, 3, 12, 44, 24, 1, 24, 42, 44, 16, 12, 42, 22, 26, 10, 8, 46, 50, 20, 42, 48, 45, 43, 35, 9, 12, 22, 2, 14, 50, 16, 29, 31, 46, 20, 35, 11, 4, 32, 35, 15, 29, 16\}$$
$$v = \{98, 74, 76, 4, 12, 27, 90, 98, 100, 35, 30, 19, 75, 72, 19, 44, 5, 66, 79, 87, 79, 44, 35, 6, 82, 11, 1, 28, 95, 68, 39, 86, 68, 61, 44, 97, 83, 2, 15, 49, 59, 30, 44, 40, 14, 96, 37, 84, 5, 43, 8, 32, 95, 86, 18\}$$

f_{17}. N = 60, W = 1006

$$w = \{7, 13, 47, 33, 38, 41, 3, 21, 37, 7, 32, 13, 42, 42, 23, 20, 49, 1, 20, 25, 31, 4, 8, 33, 11, 6, 3, 9, 26, 44, 39, 7, 4, 34, 25, 25, 16, 17, 46, 23, 38, 10, 5, 11, 28, 34, 47, 3, 9, 22, 17, 5, 41, 20, 33, 29, 1, 33, 16, 14\}$$

$$v = \{81, 37, 70, 64, 97, 21, 60, 9, 55, 85, 5, 33, 71, 87, 51, 100, 43, 27, 48, 17, 16, 27, 76, 61, 97, 78, 58, 46, 29, 76, 10, 11, 74, 36, 59, 30, 72, 37, 72, 100, 9, 47, 10, 73, 92, 9, 52, 56, 69, 30, 61, 20, 66, 70, 46, 16, 43, 60, 33, 84\}$$

f_{18}. N = 65, W = 1319

$$w = \{47, 27, 24, 27, 17, 17, 50, 24, 38, 34, 40, 14, 15, 36, 10, 42, 9, 48, 37, 7, 43, 47, 29, 20, 23, 36, 14, 2, 48, 50, 39, 50, 25, 7, 24, 38, 34, 44, 38, 31, 14, 17, 42, 20, 5, 44, 22, 9, 1, 33, 19, 19, 23, 26, 16, 24, 1, 9, 16, 38, 30, 36, 41, 43, 6\}$$

$$v = \{47, 63, 81, 57, 3, 80, 28, 83, 69, 61, 39, 7, 100, 67, 23, 10, 25, 91, 22, 48, 91, 20, 45, 62, 60, 67, 27, 43, 80, 94, 47, 31, 44, 31, 28, 14, 17, 50, 9, 93, 15, 17, 72, 68, 36, 10, 1, 38, 79, 45, 10, 81, 66, 46, 54, 53, 63, 65, 20, 81, 20, 42, 24, 28, 1\}$$

f_{19}. N = 70, W = 1426

$$w = \{4, 16, 16, 2, 9, 44, 33, 43, 14, 45, 11, 49, 21, 12, 41, 19, 26, 38, 42, 20, 5, 14, 40, 47, 29, 47, 30, 50, 39, 10, 26, 33, 44, 31, 50, 7, 15, 24, 7, 12, 10, 34, 17, 40, 28, 12, 35, 3, 29, 50, 19, 28, 47, 13, 42, 9, 44, 14, 43, 41, 10, 49, 13, 39, 41, 25, 46, 6, 7, 43\}$$

$$v = \{66, 76, 71, 61, 4, 20, 34, 65, 22, 8, 99, 21, 99, 62, 25, 52, 72, 26, 12, 55, 22, 32, 98, 31, 95, 42, 2, 32, 16, 100, 46, 55, 27, 89, 11, 8, 3, 43, 93, 53, 88, 36, 41, 60, 92, 14, 5, 41, 60, 92, 30, 55, 79, 33, 10, 45, 3, 68, 12, 20, 54, 63, 38, 61, 85, 71, 40, 58, 25, 73, 35\}$$

f_{20}. N = 75, W = 1433

$$w = \{24, 45, 15, 40, 9, 37, 13, 5, 43, 35, 48, 50, 27, 46, 24, 45, 2, 7, 38, 20, 20, 31, 2, 20, 3, 35, 27, 4, 21, 22, 33, 11, 5, 24, 37, 31, 46, 13, 12, 12, 41, 36, 44, 36, 34, 22, 29, 50, 48, 17, 8, 21, 28, 2, 44, 45, 25, 11, 37, 35, 24, 9, 40, 45, 8, 47, 1, 22, 1, 12, 36, 35, 14, 17, 5\}$$

$$v = \{2, 73, 82, 12, 49, 35, 78, 29, 83, 18, 87, 93, 20, 6, 55, 1, 83, 91, 71, 25, 59, 94, 90, 61, 80, 84, 57, 1, 26, 44, 44, 88, 7, 34, 18, 25, 73, 29, 24, 14, 23, 82, 38, 67, 94, 43, 61, 97, 37, 67, 32, 89, 30, 30, 91, 50, 21, 3, 18, 31, 97, 79, 68, 85, 43, 71, 49, 83, 44, 86, 1, 100, 28, 4, 16\}$$

References

1. Hernández, P.R., Dimopoulos, N.J.: A new heuristic for solving the multichoice multidimensional knapsack problem. Proc. IEEE Trans. Syst. Man Cybernet. Part A: Syst. Hum. **35**(5), 708–717 (2005)
2. Martello, S., Toth, P.: Knapsack Problems: Algorithms and Computer Implementations, pp. 1–296. Wiley, New York (1990)
3. Tavares, J., Pereira, F.B., Costa, E.: Multidimensional knapsack problem: a fitness landscape analysis. Proc. IEEE Trans. Syst. Man Cybernet. Part B: Syst. Hum. **38**(3), 604–616 (2008)
4. Moser, M.: Declarative scheduling for optimally graceful QoS degradation. In: Proceedings of IEEE International Conference Multimedia Computing Systems, pp. 86–93 (1996)
5. Khan, M.S.: Quality adaptation in a multisession multimedia system: model, algorithms and architecture. Ph.D. dissertation, Department of Electrical and Computer Engineering, University of Victoria, Victoria, BC, Canada (1998)
6. Warner, D., Prawda, J.: A mathematical programming model for scheduling nursing personnel in a hospital. Manag. Sci. (Appl. Ser. Part 1) **19**(4), 411–422 (1972)
7. Psinger D.: Algorithms for Knapsack Problems, PhD thesis, Department of Computer Science, University of Copenhagen, Copenhagen,Denmark, (1995)
8. Laporte, G.: The vehicle routing problem: an overview of exact and approximate algorithms. Eur. J. Oper. Res. **59**, 345–358 (1992)
9. Granmo, O.C., Oommen, B.J., Myrer, S.A., Olsen, M.G.: Learning automata-based solutions to the nonlinear fractional knapsack problem with applications to optimal resource allocation. Proc. IEEE Trans. Syst. Man Cybernet. Part B Cybernet. **37**(1), 166–175 (2007)
10. Zou, D., Gao, L., Li, S., Wu, J.: Solving 0–1 knapsack problem by a novel global harmony search algorithm. Appl. Soft Comput. **11**, 1556–1564 (2011)
11. Layeb, A.: A hybrid quantum inspired harmony search algorithm for 0–1 optimization problems. J. Comput. Appl. Math. **253**, 14–25 (2013)
12. Layeb, A.: A novel quantum inspired cuckoo search for knapsack problems. Int. J. Bio-Inspir. Comput. **3**(5), 297–305 (2011)
13. Geem, Z.W., Kim, J.H., Loganathan, G.V.: A new heuristic optimization algorithm: harmony search. Simulation **76**(2), 60–68 (2001)
14. Mahdavi, M., Fesanghary, M., Damangir, E.: An improved harmony search algorithm for solving optimization problems. Appl. Math. Comput. **188**, 1567–1579 (2007)
15. Deb, K.: An efficient constraint handling method for genetic algorithms. Comput. Methods Appl. Mech. Eng. **186**, 311–338 (2000)
16. Kulkarni, A.J., Tai, K.: Solving constrained optimization problems using probability collectives and a penalty function approach. Int. J. Comput. Intell. Appl. **10**(4), 445–470 (2011)
17. Kulkarni, A.J., Tai, K.: A probability collectives approach with a feasibility-based rule for constrained optimization. Appl. Comput. Intell. Soft Comput. **2011**, Article ID 980216 (2011)
18. Ma, W., Wang, M.: Zhu X: Improved particle swarm optimization based approach for bilevel programming problem-an application on supply chain model. Int. J. Mach. Learn. Cybernet. **5**(2), 281–292 (2014)
19. Chen, C.J.: Structural vibration suppression by using neural classifier with genetic algorithm. Int. J. Mach. Learn. Cybernet. **3**(3), 215–221 (2012)
20. Wang, X.Z., He, Q., Chen, D.G., Yeung, D.: A genetic algorithm for solving the inverse problem of support vector machines. Neurocomputing **68**, 225–238 (2005)
21. Wang, X.Z., He, Y.L., Dong, L.C., Zhao, H.Y.: Particle swarm optimization for determining fuzzy measures from data. Inf. Sci. **181**(19), 4230–4252 (2011)
22. Krishnasamy, G., Kulkarni, A.J., Paramesran, R.: A hybrid approach for data clustering based on modified cohort intelligence and k-means. Expert Syst. Appl. **41**, 6009–6016 (2014)

Chapter 6
Cohort Intelligence for Solving Travelling Salesman Problems

As demonstrated previously, the performance of the Cohort Intelligence Algorithm (CI) algorithm was quite satisfactory for solving combinatorial problem such as Knapsack Problem (KP) [1]. The purpose of this chapter is to further demonstrate:

1. The ability of the CI methodology solving classic NP-hard combinatorial problem such as the Traveling Salesman Problem (TSP). In all 9 small sized test cases (14 to 29 cities) of the TSP from the TSPLIB [2] were solved.
2. In addition to the original CI approach incorporated with a *roulette wheel selection approach* [3], two different approaches such as a best *behavior selection approach* and a *random behavior selection approach* have been incorporated. In the best behavior selection approach every candidate follows the best behavior in the cohort. In the latter approach candidates randomly select any behavior in the cohort.
3. In order to jump out of possible local minima and further make the cohort saturate to global minimum, a generic approach of perturbation and further accepting worst behaviors was successfully incorporated.

The results highlighted the simplicity of the CI algorithm as well as robustness of the solution with the three approaches. It also underscored that the CI incorporated with the roulette wheel selection approach more realistically resembles the competitive and interactive learning behavior of the cohort candidates, which eventually makes the cohort successful. In addition, it also demonstrated that always following the best behavior/solution may make the cohort to saturate faster; however may make the cohort stuck into local minima. The encouraging results may help solve the real world problems with increasing complexity as the TSP can be further generalized to a wide variety of routing and scheduling problems [4].

A.J. Kulkarni et al., *Cohort Intelligence: A Socio-inspired Optimization Method*, Intelligent Systems Reference Library 114, DOI 10.1007/978-3-319-44254-9_6

6.1 Traveling Salesman Problem (TSP)

The Travelling Salesman Problem (TSP) is a classic combinatorial NP Hard problem [5–8]. It includes N cities and one salesman. One of the N cities is considered as origin city. The salesman must start from origin city, visit all the remaining cities exactly once and must return to the origin city. The goal is to find the minimum cost (distance, time, etc.) of route/path that the salesman should follow. It is represented as $f(\mathbf{v}) = f(v_O, \ldots v_i, \ldots v_N, v_O)$ where, v_O is the origin city and v_i represents any intermediate city and $\mathbf{v} = (v_O, \ldots v_i, \ldots v_N, v_O)$ is the route the salesman follows or the order in which the salesman visits the N cities.

6.1.1 Solution to TSP Using CI

In the context of CI algorithm presented in Chap. 2 the edges in the route $\mathbf{v} = (v_O, \ldots v_i, \ldots v_N, v_O)$ are considered as the characteristics/qualities which decide the overall cost of the route. The procedure begins with the initialization of number of cohort candidates C, number of variations t and the route $\mathbf{v}^c$ of every candidate $c, \ (c = 1, \ldots, C)$.

In the cohort of C candidates, every individual candidate $c, \ (c = 1, \ldots, C)$ has a route $\mathbf{v}^c = (v_O^c, \ldots v_i^c, \ldots v_N^c, v_O^c)$ which is the order in which the candidate visits the N cities. The origin city v_O^c of the route of every candidate $c, \ (c = 1, \ldots, C)$ was fixed. The remaining cities were arranged randomly in the route. This way C routes $(\mathbf{v}^1, \ldots, \mathbf{v}^c, \ldots, \mathbf{v}^C)$ are formed. And associated costs are calculated $\mathbf{F}^C = \{f(\mathbf{v}^1), \ldots, f(\mathbf{v}^c), \ldots, f(\mathbf{v}^C)\}$. The following procedure is explained in the context of the *roulette wheel selection* approach.

(a) In the context of roulette wheel approach, the probability p^c of selecting the route $\mathbf{v}^c$ of every associated candidate $c, \ (c = 1, \ldots, C)$ is calculated.
(b) Every candidate $c, \ (c = 1, \ldots, C)$ using roulette wheel approach selects to follow a certain route $\mathbf{v}^{(c)}$ of some other candidate (c), i.e. it incorporates an edge from within $\mathbf{v}^{(c)} = \left(v_O^{(c)}, \ldots v_i^{(c)}, \ldots v_N^{(c)}, v_O^{(c)}\right)$ and incorporates into its existing route $\mathbf{v}^c = (v_O^c, \ldots v_i^c, \ldots v_N^c, v_O^c)$. Following a route $\mathbf{v}^{(c)} = \left(v_O^{(c)}, \ldots v_i^{(c)}, \ldots v_N^{(c)}, v_O^{(c)}\right)$ means incorporating an edge from it into its existing route $\mathbf{v}^c = (v_O^c, \ldots v_i^c, \ldots v_N^c, v_O^c)$. More specifically, an edge $(v_i^{(c)}, v_{i+1}^{(c)})$ from within $\mathbf{v}^{(c)} = \left(v_O, \ldots v_i^{(c)}, \ldots v_N^{(c)}, v_O\right)$ is selected. The positions of the cities of this edge in the route $\mathbf{v}^c = (v_O, \ldots v_i^c, \ldots v_N^c, v_O)$ of candidate c are identified. Then the position of the city which is farther from the city of origin v_O^c in the route is swapped with the position of the city immediately after the city which is closer to origin v_O^c. In other words, the positions of the cities of this edge

$\left(\text{i.e. } \left(v_i^{(c)}, v_{i+1}^{(c)}\right)\right)$ in the route $\mathbf{v}^c = (v_O, \ldots v_i^c, \ldots v_N^c, v_O)$ of candidate c are identified (say a and b). Then cities at positions $a + 1$ and b are swapped, which makes the edge $\left(v_i^{(c)}, v_{i+1}^{(c)}\right)$ a part of the route $\mathbf{v}^c$.

(c) In this way, every candidate c, $(c = 1, \ldots, C)$ forms t new independent routes with associated costs $\mathbf{F}^{c,t} = \left\{f(\mathbf{v}^c)^1, \ldots, f(\mathbf{v}^c)^j, \ldots, f(\mathbf{v}^c)^t\right\}$, $(c = 1, \ldots, C)$ and further selects the best cost route amongst them. This makes the cohort available with C updated routes with their costs represented as $\mathbf{F}^C = \{f^*(\mathbf{v}^1), \ldots, f^*(\mathbf{v}^c), \ldots, f^*(\mathbf{v}^C)\}$.

This process continues until saturation, i.e. every candidate finds the same route and does not change for successive considerable number of learning attempts. The above discussed procedure of solving TSP using CI algorithm is illustrated in Figs. 6.1, 6.2 and 6.3 with a 5 city TSP, $C = 3$, origin city v_O as 1, the corresponding route $\mathbf{v}^c$.

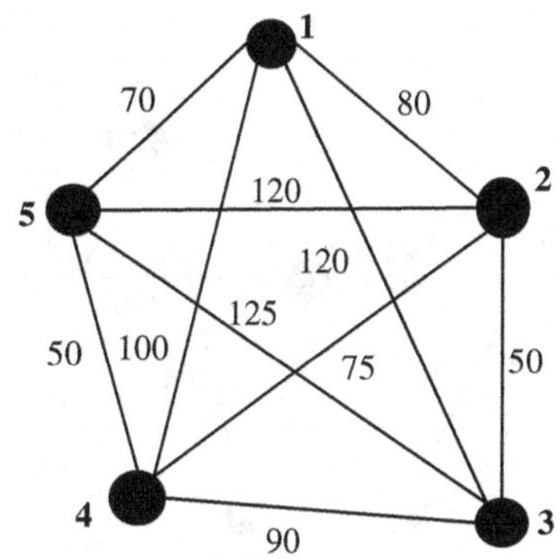

$$\begin{bmatrix} - & 80 & 120 & 100 & 70 \\ 80 & - & 50 & 75 & 120 \\ 120 & 50 & - & 90 & 125 \\ 100 & 75 & 90 & - & 50 \\ 70 & 120 & 125 & 50 & - \end{bmatrix}$$

Cost matrix

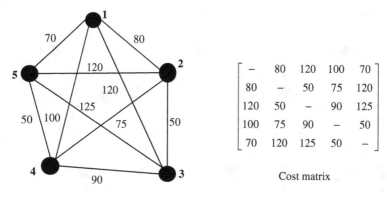

Fig. 6.1 Illustrative example with 5 cities

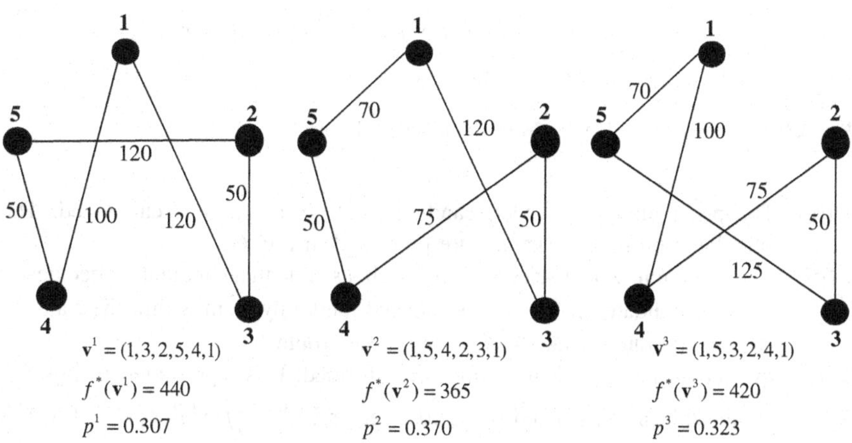

Fig. 6.2 Illustrative example with 5 cities (candidate solutions)

Route of candidate 1: $\mathbf{v}^1 = (1,3,2,4,5,1)$, Route of candidate 2: $\mathbf{v}^2 = (1,5,4,2,3,1)$

$t=1$: edge sampled from within $\mathbf{v}^2$ using roulette wheel selection: $(1,5)$

The cities are identified in $\mathbf{v}^1$: $(\boxed{1},3,2,\boxed{5},4,1)$

Update the route $\mathbf{v}^1$ by incorporating the edge $(1,5)$: $(1,\boxed{5},2,\boxed{3},4,1)$

Cost of route $f(\mathbf{v}^1)^1$: 430

$t=2$: edge sampled from within $\mathbf{v}^2$ using roulette wheel selection: $(4,2)$

The cities are identified in $\mathbf{v}^1$: $(1,3,\boxed{5},4,\boxed{2},1)$

Update the route $\mathbf{v}^1$ by incorporating the edge $(4,2)$: $(1,3,2,\boxed{4},\boxed{5},1)$

Cost of route $f(\mathbf{v}^1)^2$: 375

$t=3$: edge sampled from within $\mathbf{v}^2$ using roulette wheel selection: $(5,4)$

The cities are found: $(1,3,2,\boxed{5},\boxed{4},1)$

No changes in the route by including edge $(5,4)$: $(1,3,2,5,4,1)$

Cost of route $f(\mathbf{v}^1)^3$: 440

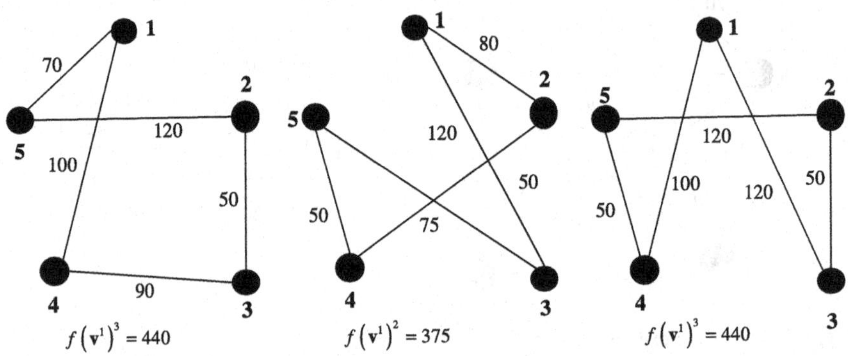

$$\mathbf{F}^{1,3} = \left\{ f(\mathbf{v}^1)^1, f(\mathbf{v}^1)^2, f(\mathbf{v}^1)^3 \right\} = \{430, 375, 440\}$$

$$f^*(\mathbf{v}^1) = 375$$

Fig. 6.3 Illustrative example with 5 cities (variations obtained)

3.1(a) The probability p^c of each candidate $c, (c = 1, \ldots, 3)$ is calculated. The calculated probability values are presented in Fig. 6.2.

3.1(b) Using roulette wheel selection approach, assume that candidate 1 decides to follow candidate 2. An edge is selected randomly from within the route $\mathbf{v}^2$ and incorporated it into $\mathbf{v}^1$ forming a new route.

3.1(c) In such manner $t = 3$ new routes are formed. It is represented in Fig. 6.3 along with the associated route cost vector $\mathbf{F}^{1,3} = \left\{ f(\mathbf{v}^1)^1, f(\mathbf{v}^1)^2, f(\mathbf{v}^1)^3 \right\}$

and the selected best route with cost $f^*(\mathbf{v}^1)$. It is presented in Fig. 6.3 In this way, candidates 2 and 3 also follow certain candidate's route and update their own routes. It makes the cohort available with 3 updated routes with costs $\mathbf{F}^3 = \{f^*(\mathbf{v}^1), f^*(\mathbf{v}^2), f^*(\mathbf{v}^3)\}$.

This process continues until saturation (convergence) i.e. every candidate finds the same route and does not change for successive considerable number of learning attempts.

In the context of the illustration provided in Figs. 6.1 and 6.2, in case of the best behavior selection approach every candidate in the cohort will follow candidate 2 as it has the minimum cost the current learning attempt. And the in case of the random behavior selection approach, every candidate selects a candidate randomly and follows its behavior. The results of the CI approach solving the TSP are discussed in the next section.

6.2 Results and Discussion

The CI algorithm discussed in Chap. 2 applied for solving the TSP was coded in MATLAB 7.7.0 (R2008b) and simulations were run on a Windows platform using i3-M380, 2.53 GHz processor speed with 3 GB RAM. The number of candidates C and variations t were chosen to be 5 and 5, respectively. In all, nine cases of the TSP [2] with number of cities varying from 14 to 29 were solved. Every case was solved 20 times. In the earlier version of CI [3], the CI candidates used roulette wheel approach for the selection of the behaviour in the cohort to follow. In addition to it, the CI algorithm here was successfully implemented with the best behavior selection and random behavior selection approach. The results are summarized in Table 6.1 with representative saturation history plots of the cohort with 5 candidates are presented in Fig. 6.4a–c.

It could be understood that, the cohort with roulette wheel selection approach saturated/converged in every run solving every case of the TSP to a marginally better solution than the best behaviour and random behaviour selection approach. In addition, even though the computational cost (time and function evaluations) of the CI with roulette wheel approach was more than CI with the best behaviour approach incorporated, the standard deviation (SD) exhibited comparatively more robustness (refer to Fig. 6.5b). The inherent probabilistic nature of the roulette wheel selection approach helped the algorithm explore a better solution from within the cohort. Also in some of the runs, similar to the feasibility-based rule developed in [9, 10], this approach necessarily helped the CI candidates jump out of local minima by following worse behaviour. It is important to mention here that the overall tendency to improve by competition and interaction ensured the saturation to an optimal solution. Furthermore, as presented in Table 6.1 and Fig. 6.5a, c the best behavior selection approach was found to be computationally (time and function evaluations) cheaper; however, the cohort stuck into local minima and did not yield a better

Table 6.1 Summary of CI performance solving the TSP

Problem name	Cities (N)	Reported optimum [2]	Behavior selection approaches	Cohort candidates' saturated/converged route costs	Closeness to reported optimum (%)	Average function evaluations (FE)	Time (sec)	Accepted route cost $f^*(\mathbf{v})$			SD
								Best	Mean	Worst	
Burma14	14	30.8785	Best	[31.2321, 31.2321, 31.2321, 31.2321, 31.2321]	1.1451	105,000	2.7	31.2321	33.2283	35.6267	2.225
			Roulette	[30.8785, 30.8785, 30.8785, 30.8785, 30.8785]	0.0000	1,575,000	30.88	30.8785	30.8785	30.8785	0.00
			Random	[31.3712, 31.3712, 31.3712, 31.3712, 31.3712]	1.5956	2,625,000	102.1	31.3712	31.9577	32.5442	0.830
P01	15	284.3809	Best	[363.295, 363.295, 363.295, 363.295, 363.295]	9.7578	120,000	3.8	363.295	352.9675	342.640	14.605
			Roulette	[284.381, 284.381, 284.381, 284.381, 284.381]	0.0000	1,320,000	41.67	284.381	284.381	284.381	0.00
			Random[a]	[284.381, 284.381, 360.830, 323.392, 323.392]	0.0000	3,000,000	69.36	284.381	284.381	284.381	0.00
Ulysses16	16	74.1087	Best	[84.0071, 84.0071, 84.0071, 84.0071, 84.0071]	13.3565	120,000	4.02	84.0071	84.2699	84.5327	0.372
			Roulette	[73.9876, 73.9876, 73.9876, 73.9876, 73.9876]	0.1634	2,400,000	85.50	73.9876	74.026	74.3079	0.099
			Random[a]	[83.4726, 79.5323, 76.9698, 80.4994, 77.2778]	3.8606	3,000,000	102.7	76.9698	77.7545	78.5392	1.110

(continued)

Table 6.1 (continued)

Problem name	Cities (N)	Reported optimum [2]	Behavior selection approaches	Cohort candidates' saturated/converged route costs	Closeness to reported optimum (%)	Average function evaluations (FE)	Time (sec)	Accepted route cost $f^*(\mathbf{v})$			SD
								Best	Mean	Worst	
Groetschel17	17	2085	Best	[2311, 2311, 2311, 2311, 2311]	10.8393	675,000	19.22	2311	2342	2373	43.84
			Roulette	[2085, 2085, 2085, 2085, 2085]	0.0000	2,025,000	70.74	2085	2086.4	2090	1.90
			Random	[2245, 2245, 2245, 2245, 2245]	7.6738	3,375,000	114.2	2245	2614.5	2984	522.6
Groetschel21	21	2707	Best	[3182, 3182, 3182, 3182, 3182]	17.5474	1,365,000	40.28	3182	3381	3580	281.4
			Roulette	[2707, 2707, 2707, 2707, 2707]	0.0000	2,100,000	72.16	2707	2707.6	2709	0.98
			Random	[2796, 2796, 2796, 2796, 2796]	32.8777	2,415,000	80.23	2796	2950.5	3105	218.5
Ulysses22	22	75.5975	Best	[96.1869, 96.1869, 96.1869, 96.1869, 96.1869]	27.2355	160,000	6.92	96.1869	97.895	99.6095	2.42
			Roulette	[75.5975, 75.5975, 75.5975, 75.5975, 75.5975]	0.0000	3,520,000	120.3	75.5975	76.607	77.8509	0.37
			Random[a]	[99.0993, 76.5557, 88.3879, 76.5557, 86.5087]	1.2675	3,520,000	121.4	76.5557	77.995	79.4341	2.04
Groetschel24	24	1272	Best	[1391, 1391, 1391, 1391, 1391]	9.3553	4,140,000	66.62	1391	1492	1467	53.74
			Roulette	[1272, 1272, 1272, 1272, 1272]	0.0000	4,320,000	70.42	1272	1282.6	1318	14.85
			Random	[1429, 1429, 1429, 1429, 1429, 1429]	12.3427	4,500,000	101.7	1429	1488.5	1548	84.14

(continued)

Table 6.1 (continued)

Problem name	Cities (N)	Reported optimum [2]	Behavior selection approaches	Cohort candidates' saturated/converged route costs	Closeness to reported optimum (%)	Average function evaluations (FE)	Time (sec)	Accepted route cost $f^*(\mathbf{v})$			SD
								Best	Mean	Worst	
Fri26	26	937	Best	[1156, 1156, 1156, 1156, 1156]	23.3724	5,062,500	154.3	1156	1180	1204	33.94
			Roulette	[937, 937, 937, 937, 937]	0.0000	5,062,500	154.3	937	941.6	959	8.38
			Random	[1085, 1085, 1085, 1085, 1085]	15.7951	5,062,500	154.3	1085	1210	1335	176.8
Bays29	29	9074	Best solution	[12665, 12665, 12665, 12665, 12665]	39.5746	350,000	13.65	12665	13013	13361	492.1
			Roulette	[9108.8, 9108.8, 9108.8, 9108.8, 9108.8]	0.3835	3,500,000	143.6	9108.8	9424.56	9654.3	192.26
			Random	[9122, 9122, 9122, 9122, 9122]	0.5290	5,250,000	200.2	9122	9341.5	9561	310.4

[a]Cohort failed to saturate

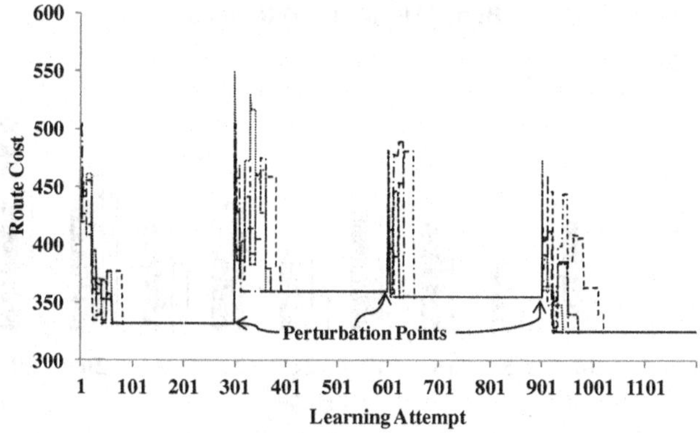

(a) **CI Saturation (Best Behavior Selection)**

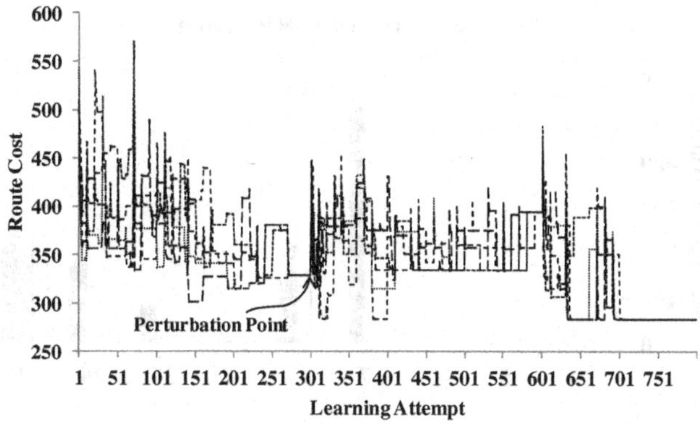

(b) **CI Saturation (Random Behavior Selection)**

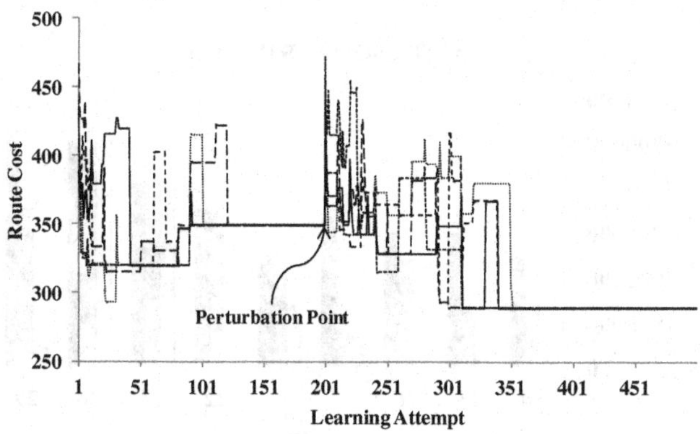

(c) **CI Saturation (Roulette Behavior Selection)**

Fig. 6.4 CI saturation history solving the TSP (P01, 5 candidates)

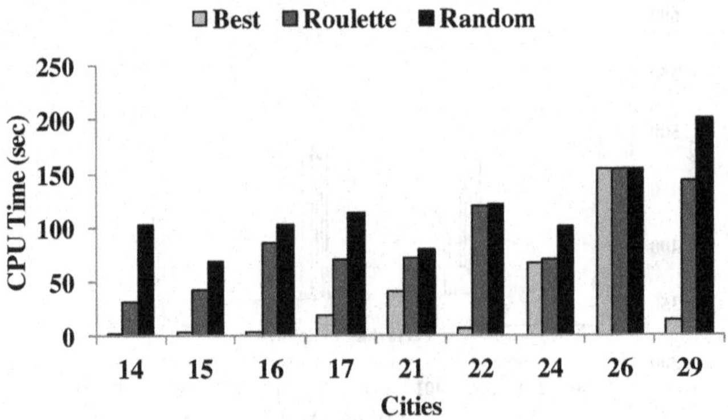

(a) CI CPU Time Performance

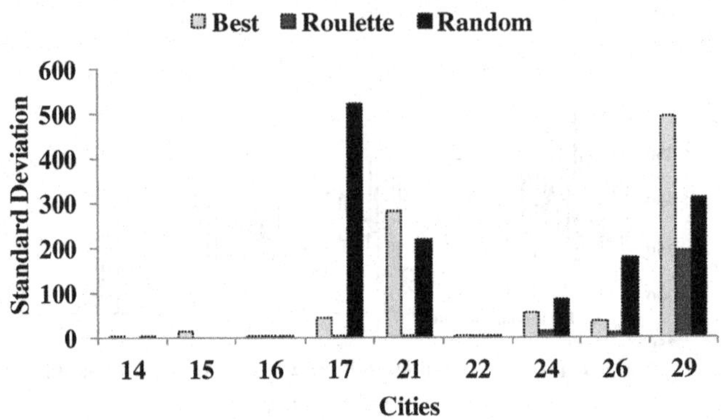

(b) CI Standard Deviation

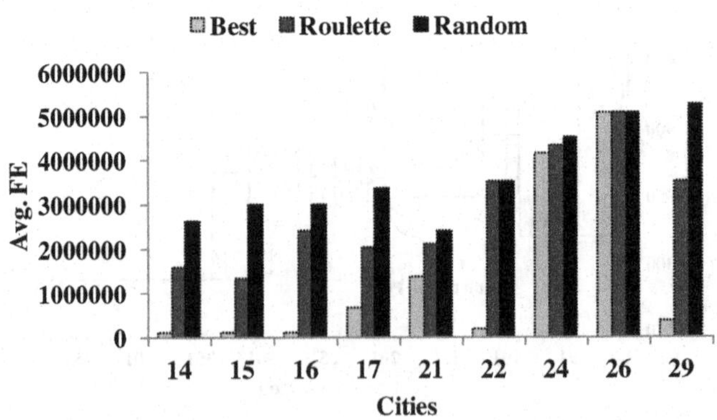

(c) CI Average Function Evaluations

Fig. 6.5 Illustration of CI performance

solution. This happened because all the candidates in the cohort followed the best behaviour in every learning attempt, and the variation necessary for exploration and avoidance of the local minima was restricted. In the CI with random selection approach, following any candidate randomly forced the candidates explore the larger search space; however resulted into extremely slow saturation of the cohort behavior. Moreover as exhibited in Table 6.1, the saturation was not achieved in every run of the algorithm. The SD presented in Table 6.1 and Fig. 6.5a–c indicated that for solving all the cases the CI with roulette wheel selection approach was comparatively more robust. However, as the problem size increased, robustness of all the approaches was reduced. It was also observed that the computational cost of all the approaches increased with the increase in problem size.

Once the cohort behavior was saturated every candidate perturbed its individual solution. More specifically, the routes of all the candidates were randomly altered by changing the order of a certain number of cities in the vector $\mathbf{v}^c = \left(v_O^c, \ldots v_i^c, \ldots v_n^c, v_O^c\right)$ for every candidate c, $(c = 1, \ldots, N)$. It is important to mention here that the perturbation approach in [9, 10] required several parameters to be tuned which was completely avoided here. As evident in Fig. 6.4a–c, this approach was found to be helping the individual candidate's solution jump out of local minima and further saturates the cohort behavior to a significantly improved solution.

6.3 Concluding Remarks and Future Directions

For the first time emerging CI algorithm has been applied for solving combinatorial NP-hard problem such as the TSP with number of cities varying from 14 to 29. The rational and self supervising learning nature of the cohort candidates was successfully formulated and demonstrated along with the learning and improving qualities which further improved their individual behavior. The application of the CI methodology for solving combinatorial NP-hard problem such as the TSP is successfully demonstrated. The CI incorporated with the roulette wheel approach, best behavior selection as well as random behavior selection approaches was successfully presented. The results highlighted the overall simplicity of the algorithm as well as robustness of the solution with the roulette wheel approach. It also underscored that the CI incorporated with the roulette wheel selection approach more realistically resembles the competitive and interactive learning behavior of the cohort candidates, which eventually drove the cohort to marginally improved solution. Moreover, it is also demonstrated that always following the best behavior/solution may make the cohort to saturate faster; however may make the cohort stuck into local minima. In addition, in order to jump out of possible local minima and further make the cohort saturate to global minimum, a generic approach such as accepting worst behaviors was incorporated. The encouraging results may help solve the real world problems with increasing complexity as the TSP can be further generalized to a wide variety of routing and scheduling problems [4]. In addition, CI approach could be modified to make it solve Multiple TSP (MTSP) and Vehicle Routing Problem (VRP).

In addition to the advantages few limitations are also observed. A generic fine parameter tuning approach needs to be developed for selection of the parameters such as number of candidates C and number of variations t. In this chapter we have solved problems of sizes up to 29 cities. As the problem size was increased, the solutions obtained were less robust and the global minimum was found only intermittently. A possible solution to this problem is to develop a distributed CI approach wherein larger sized problems could be decomposed into smaller size problems and solve them independently. In this context, author see potential real world applications related to the distributed communication system such as, path planning of Unmanned Aerial vehicles (UAV) and addressing the ever growing traffic control problem using Vehicular ad hoc network (VANET).

References

1. Kulkarni, A.J., Shabir, H.: Solving 0–1 knapsack problem using cohort intelligence algorithm. Int. J. Mach. Learn. Cybern. (2014). doi 10.1007/s13042-014-0272-y
2. Reinelt, G.: TSPLIB—a travelling salesman problem library. ORSA J. Comput. **3**, 376–384 (1991)
3. Kulkarni, A.J., Durugkar, I.P., Kumar, M.: Cohort intelligence: a self-supervised learning behavior. In: IEEE International Conference on Systems, Man, and Cybernetics, Manchester, UK, pp. 1396–1400 (2013)
4. Somhom, S., Modares, A., Enkawa, T.: Competition-based neural network for the multiple travelling salesman problem with minmax objective. Comput. Oper. Res. **26**(4), 395–407 (1999)
5. Calvo, R.W., Cordone, R.: A heuristic approach to the overnight security service problem. Comput. Oper. Res. **30**(9), 1269–1287 (2003)
6. Christofides, N., Eilon, S.: An algorithm for the vehicle dispatching problem. Oper. Res. Q. **20**(3), 309–318 (1969)
7. Jati, G.K., Manurung, H.M., Suyanto, S.: Discrete cuckoo search for travelling salesman problem. In: IEEE International Conference on Computing and Convergence Technology, Seoul, Korea, pp. 993–997 (2012)
8. Luciana, B., Frana, P.M., Pablo, M.: A new memetic algorithm for the asymmetric travelling salesman problem. J. Heuristics **10**(5), 483–506 (2004)
9. Kulkarni, A.J., Tai, K.: A probability collectives approach with a feasibility-based rule for constrained optimization. Appl. Comput. Intell. Soft Comput. (2011), Article ID 980216
10. Kulkarni, A.J., Tai, K.: A probability collectives approach for multi-agent distributed and cooperative optimization with tolerance for agent failure. In: Czarnowski, I., Jedrzejowicz, P., Kacprzyk, J. (eds.) Agent Based Optimization, Studies in Computational Intelligence, vol. 456, pp. 175–201 Springer, Berlin-Heidelberg (2013)

Chapter 7
Solution to a New Variant of the Assignment Problem Using Cohort Intelligence Algorithm

Nomenclature

C An n by n row circular matrix. The ijth element of C is $C_{i,j}$, where $i,j = 1,\ldots,n$

N The set of integers $\{1,2,\ldots,n\}$

π A permutation of set N

C^{π} An n by n matrix obtained by shifting each element of row i of matrix C for $i = 1,\ldots,n$ by $(\pi(i) - 1)$ positions to the right in a circular manner. In other words, The ikth element of C^{π} is given by $C^{\pi}_{i,k} = C_{i,k-\pi(i)+1} \forall\ 1 \le i \le n$, $1 \le k \le n$

I_k The sum of the kth column of matrix C^{π}

Z The maximum column sum of matrix C^{π}, $Z = max_{k=1}^{n}\{I_k\}$

x_{ij} A binary variable equal to 1 if $\pi(i) = j$; and 0, otherwise

In this chapter, we present a variant of the classical assignment problem [1]. The model has applications in healthcare systems and inventory management. The problem stems from an application in healthcare management. Specifically, a surgical scheduling in a hospital setting is a complex combinatorial problem. In addition, similar problem arises in minimizing the space requirements in a retail store. The problem formulation, applications and solution using Cohort Intelligence methodology [2–4] is presented in sections below.

7.1 New Variant of the Assignment Problem

Suppose we seek to schedule n surgeons/doctors over a planning horizon of n days. The recovery time for each operated patient in the recovery room varies according to the type of surgery. When building cyclic surgery schedules, one important objective is to minimize congestion in the recovery room. That is, we want to minimize the maximum number of patients in the recovery unit in any given day of

A.J. Kulkarni et al., *Cohort Intelligence: A Socio-inspired Optimization Method*, Intelligent Systems Reference Library 114, DOI 10.1007/978-3-319-44254-9_7

the planning horizon so that the costs associated with important resources such as nurses, space, beds, and equipment are also minimized. Another application of this problem arises in supply chain management. By considering cyclic scheduling for suppliers, the maximum required storage space of a retail shop on any day over a planning horizon of n days can be minimized by developing optimal delivery schedules. The mathematical statement and formulation of the problem are discussed below in detail.

As in [5] a row vector is said to be circular if its first and last elements are considered to be consecutive. A matrix is called row circular if its rows are circular. Given an $(n \times n)$ row circular matrix $C = \{C_{i,j}\}$, the problem is to minimize

$$Z = \max_{k=1}^{n} \sum_{i=1}^{n} C_{i,k-\pi(i)+1}$$

where $\pi = (\pi(1), \pi(2), \ldots, \pi(n))$ is a permutation of the set $N \equiv \{1, 2, \ldots, n\}$. Matrix C being row circular implies that $C_{i,j\pm n} = C_{i,j} \forall i, j$. We call this problem a Cyclic Bottleneck Assignment Problem (CBAP). Cyclic refers to the row circularity of matrix C; bottleneck refers to the *min max* objective; and assignment refers to the problem's close affinity to the classical assignment problem that minimizes $\sum_{i=1}^{n} C_{i,\pi(i)}$.

To give the problem a different description, for a given permutation π of the set N, let's define matrix C^{π} by moving each element of row $i, i = 1, \ldots, n$, of matrix C by $(\pi(i) - 1)$ positions to the right in a circular manner. More precisely, let $C^{\pi}_{i,k} = C_{i,k-\pi(i)+1} \forall\ 1 \le i \le n, 1 \le k \le n$. Since π is a permutation, every row of matrix C is rotated by a different number of columns to obtain the rotated matrix C^{π}. Furthermore, let I_k denote the sum of the kth column of the rotated matrix C^{π}. In other words, let $I_k = \sum_{i=1}^{n} C^{\pi}_{i,k} = \sum_{i=1}^{n} C_{i,k-\pi(i)+1}$. With these new terms, the objective in our problem can be stated as min $max_{k=1}^{n}\{I_k\}$. That is, the problem is to find a permutation that minimizes the maximum column sums of the rotated matrix. Note that, with the above notation, the standard assignment problem is equivalent to min $min_{k=1}^{n} \sum_{i=1}^{n} C^{\pi}_{i,k}$.

To formulate the integer linear programming model for this problem, we define the following decision variables:

$$x_{i,j} = \begin{cases} 1 & \text{if } j = \pi(i) \\ 0 & \text{otherwise} \end{cases}$$

The model is given by

$$\begin{aligned} &\textit{Minimize } Z \\ &\textit{Subject to} \end{aligned} \tag{7.1}$$

$$\sum_{i=1}^{n} x_{i,j} = 1, \quad \forall\ 1 \le j \le n \tag{7.2}$$

$$\sum_{j=1}^{n} x_{i,j} = 1, \quad \forall\ 1 \le i \le n \tag{7.3}$$

$$I_k = \sum_{i=1}^{n}\sum_{j=1}^{n} C_{i,k-j+1}x_{i,j} = \sum_{i=1}^{n}\sum_{j=1}^{k} C_{i,k-j+1}x_{i,j} + \sum_{i=1}^{n}\sum_{j=1}^{n} C_{i,k-j+1+n}x_{i,j}, \quad \forall\ 1 \le k \le n \tag{7.4}$$

$$Z \ge I_k, \quad \forall\ 1 \le k \le n \tag{7.5}$$

$$x_{i,j} \in \{0,1\}, \quad \forall\ 1 \le i \le n, 1 \le j \le n \tag{7.6}$$

The objective function in Eq. 7.1 minimizes the maximum column sum of the rotated matrix C^{π}. Constraint 7.2 ensures that for each j there exists an i such that $j = \pi(i)$. Constraint 7.3 ensures that for each i there exists a j such that $j = \pi(i)$. Constraints 7.4 computes the sum of the kth column of the rotated matrix C^{π}. Constraint 7.5 sets the value of the objective function equal to the maximum column sum of the rotated matrix C^{π}. The CBAP is an NP-hard problem. For the proof of NP-hardness refer to the Appendix B provided in [1].

As an illustrative example, consider the following (3×3) row circular matrix:

$$C = \begin{bmatrix} 6 & 4 & 2 \\ 8 & 8 & 8 \\ 7 & 7 & 0 \end{bmatrix}$$

Applying the two permutations $\pi_1 = (1,2,3)$ and $\pi_2 = (1,3,2)$ of the set $\{1,2,3\}$ to matrix C yields the following rotated matrices:

$$C^{(1,2,3)} = \begin{bmatrix} 6 & 4 & 2 \\ 8 & 8 & 8 \\ 7 & 0 & 7 \end{bmatrix}, \quad C^{(1,3,2)} = \begin{bmatrix} 6 & 4 & 2 \\ 8 & 8 & 8 \\ 0 & 7 & 7 \end{bmatrix}$$

Since the column sums corresponding to permutations π_1 and π_2 are 21, 12, 17 and 14, 19, 17, respectively, the optimal solution is given by permutation π_2 yielding a minimum Z value of 19. Note that due to the row circularity, we need to consider only 2 permutations in this example and $(n-1)!$ permutations in general. While the optimal solution in this example is given by π_2, the optimal solution to the standard assignment problem is given by π_1 with a minimum objective value of 12. Before closing this section, we note that if the set of constraints given in (6.5) is replaced by the single constraint $Z \ge I_1$, then we get the classical assignment problem. To see this, define

$$\hat{x}_{i,j} = \begin{cases} x_{i,j} & \text{if } j = 1 \\ x_{i,2-j+n} & \text{if } 2 \le j \le n \end{cases}$$

Now the problem $\{(6.1\text{-}6.4), Z \ge I_1, (6.6)\}$ is equivalent to $Min \sum_{i=1}^{n} \sum_{j=1}^{n} C_{i,j}\hat{x}_{i,j}$ s.t. $\sum_{i=1}^{n} \hat{x}_{i,j} = 1 \; \forall \, j, \sum_{j=1}^{n} \hat{x}_{i,j} = 1 \; \forall \, i, \hat{x}_{i,j} \in \{0, 1\}$, which is the assignment problem.

7.2 Probable Applications

As mentioned earlier the model stated above has applications in healthcare scheduling and supply chain management. Two specific applications of this model are described below.

7.2.1 Application in Healthcare

The problem arises in surgical scheduling in a hospital setting. Surgeons operate on patients in the surgery unit. After completion of the surgery, patients are sent to the recovery unit. Assume that there are n types of surgeries that need to be performed over a planning horizon of n time periods (e.g. days). The goal is to develop a cyclic surgery schedule so as to minimize congestion in the recovery unit. Cyclic means that the schedule is repeated every n days. Also assume that the surgery unit is open every day; that exactly one type of surgery must occur in each time period; and that patients do not stay more than n days in the recovery unit. (The last assumption does not lose generality. If patients are allowed to stay more than n days in the recovery unit, an equivalent problem can be formulated in which patients stay at most n days.). For the case in which the identity permutation, $\pi(i) = i \; \forall \, i$, is the current schedule (or assignment, i.e. surgery type i is scheduled on day i), $C_{i,j}$ represents the number of patients that are operated on day i and are then sent to the recovery unit to remain there until the *end of day* $(i + j - 1)$. In general for a permutation π of the set $N \equiv \{1, 2, \ldots, n\}$, the kth column sum, I_k, of the rotated matrix C^{π} represents the number of patients remaining in the recovery unit at the end of day k. The maximum column sum of the rotated matrix C^{π} represents the maximum number of patients in the recovery unit over the planning horizon. It is desirable to keep the maximum number of patients as low as possible in order to reduce the requirement of beds, nurses and other variable costs. Then, it is reasonable to ask if there exists a different permutation that can reduce the maximum number of patients. Suppose, for example, for a given permutation π, we can find another permutation π' such that $\pi'(1) = \pi(2), \pi'(2) = \pi(1), and \; \pi'(i) = \pi(i) \; \forall \; 3 \le i \le n$ and the maximum column sum of the rotated matrix $C^{\pi'}$ is less than that of C^{π}. Then, in this case, the hospital can reduce the congestion in the recovery unit by creating a new schedule in which the positions of the surgeons that are scheduled on day 1 and day 2 are swapped and all the other surgeons keep their existing positions in the schedule. Of course, we assume that such a swap is always possible.

7.2.2 Application in Supply Chain Management

The problem arises in minimizing the space requirements in a retail store. Suppliers deliver n different types of goods on n different days, i.e. exactly one type of product is delivered per day. In this application, $C_{i,j}$ represents the amount of space required at the *beginning of day* $(i+j-1)$ for products delivered on day i. Again, we assume that suppliers deliver according to a cyclic scheduling; that the planning horizon is n days; that the retail store is open every day; and that no product stays in the store for more than n days. The identity permutation represents the current schedule, and I_k represents the space requirement at the beginning of day K. Assuming that suppliers delivering on day i can be swapped with those that make deliveries on day i' for any $1 \leq i \leq n, 1 \leq i' \leq n, i \neq i'$, the importance of the objective $\min max_{k=1}^{n} \sum_{i=1}^{n} C_{i,k}^{\pi}$ is to minimize the maximum space requirement.

7.3 Cohort Intelligence (CI) Algorithm for Solving the CBAP

The CBAP presented in Sect. 7.1 is solved using the CI algorithm discussed in Chap. 2. The adaption and implementation of CI methodology for this problem is discussed below in detail.

In the context of the CI algorithm the elements of the rearrangement/permutation vector $\pi = (\pi(1), \ldots, \pi(i), \ldots, \pi(n))$ are considered the characteristics/attributes/qualities that candidates in the cohort select and are associated with. The procedure begins with the initialization of number of cohort candidates S, number of variations Y, the permutation π^s of every candidate $s, (s = 1, \ldots, S)$ and the convergence parameter ε and maximum number of allowable learning attempts L_{max}.

In the cohort of S candidates, every individual candidate $s, (s = 1, \ldots, S)$ randomly generates a permutation $\pi^s = (\pi(1)^s, \ldots, \pi(i)^s, \ldots, \pi(n)^s)$. Every candidate s forms matrix C^{π^s} by applying its permutation π^s and rotating all the corresponding n rows of matrix C accordingly. This way, S rotated matrices $\left(C^{\pi^1}, \ldots, C^{\pi^s}, \ldots, C^{\pi^S}\right)$ are formed. Next the associated vector of maximum column sums is calculated as $Z^S = \left\{Z\left(C^{\pi^1}\right), \ldots, Z\left(C^{\pi^s}\right), \ldots, Z\left(C^{\pi^S}\right)\right\}$ where $Z\left(C^{\pi^s}\right) = max_{k=1}^{n} I_k$ and $I_k = \sum_{i=1}^{n} C_{i,k-\pi(i)^s+1}^{s}$.

Step 1. As a minimization problem, the probability P^s of selecting a column sum $Z(C^{\pi^s})$ of every candidate is calculated as follows:

$$P^s = \frac{1/Z(C^{\pi^s})}{\sum_{s=1}^{S} 1/Z(C^{\pi^s})}, \quad (s = 1, \ldots, S) \tag{7.7}$$

Step 2. Every candidate s, $(s = 1, \ldots, S)$ using a roulette wheel approach selects a candidate $\widehat{s} \in (1, \ldots, S)$ in the cohort to follow, i.e. it incorporates an element from within $\pi^{\widehat{s}}$ into its existing permutation π^s. Following a permutation means incorporating certain elements from within $\pi^{\widehat{s}}$ into π^s. More specifically, an element $\pi(i)^{\widehat{s}}$ from within $\pi^{\widehat{s}}$ is selected randomly. Then the selected element $\pi(i)^{\widehat{s}}$ is identified in π^s along with its location. It then swaps its position with the element at the location in $\pi^{\widehat{s}}$ corresponding to its current location in π^s. This way every candidate generates Y number of permutations represented as $\Pi^{s,Y} = \{\pi^{s,1}, \ldots, \pi^{s,y}, \ldots, \pi^{s,Y}\}$, $s = 1, \ldots, S$ and further computes the associated maximum column sums $Z(C^{\pi^s})^Y = \{Z(C^{\pi^s})^1, \ldots, Z(C^{\pi^s})^y, \ldots, Z(C^{\pi^s})^Y\}$, $s = 1, \ldots, S$. The minimum from within $Z(C^{\pi^s})^Y$ for every candidate s, $(s = 1, \ldots, S)$ is found along with the associated permutation.

Step 3. If either of the two criteria listed below is valid, accept any of the matrices from within the pool of current available rotated matrices $C^{\pi^s}, (s = 1, \ldots, S)$ as the saturated/converged matrix C^* and associated permutation π^* as the final solution and stop, else continue to *Step 1*.

(a) If the maximum number of learning attempts is exceeded.
(b) The cohort reaches a saturation state. There is no significant improvement in the elements of Z^S and the difference between these elements is not very significant if further learning attempts are considered. That is, the cohort saturates to the same minimum column sum for any other number of successive learning attempts.

7.3.1 A Sample Illustration of the CI Algorithm for Solving the CBAP

The CI algorithm for solving CBAP is now illustrated for the example shown in Fig. 7.1. In this example, the number of candidates is $S = 3$, the number of variations is $Y = 2$, and the number of learning attempts is $L = 1$. The initial C matrix is shown in Fig. 7.1.

1. The candidates randomly generate permutations represented as π^1, π^2, and π^3 in Fig. 7.1a. Then the corresponding rotated matrices $\left(C^{\pi^1}, C^{\pi^2}, C^{\pi^3}\right)$ and associated maximum column sums $Z^3 = \left\{Z\left(C^{\pi^1}\right), Z\left(C^{\pi^2}\right), Z\left(C^{\pi^3}\right)\right\}$ are obtained.
2. The probability P^s, $s = 1, 2, 3$ is calculated using Eq. 7.7. The calculated probability values are presented in Fig. 7.1a.

$$C = \begin{bmatrix} 40 & 30 & 22 & 13 \\ 36 & 29 & 23 & 7 \\ 32 & 30 & 22 & 7 \\ 36 & 32 & 23 & 17 \end{bmatrix}$$

$\pi^1 = (3,4,2,1)$ $\quad$ $\pi^2 = (2,3,4,1)$ $\quad$ $\pi^3 = (1,4,3,2)$

$$C^1 = \begin{bmatrix} 22 & 13 & 40 & 30 \\ 29 & 23 & 7 & 36 \\ 7 & 32 & 30 & 22 \\ 36 & 32 & 23 & 17 \end{bmatrix} \quad C^2 = \begin{bmatrix} 13 & 40 & 30 & 22 \\ 23 & 7 & 36 & 29 \\ 30 & 22 & 7 & 32 \\ 36 & 32 & 23 & 17 \end{bmatrix} \quad C^3 = \begin{bmatrix} 40 & 30 & 22 & 13 \\ 29 & 23 & 7 & 36 \\ 22 & 7 & 32 & 30 \\ 17 & 36 & 32 & 23 \end{bmatrix}$$

$Z(C^{\pi^1}) = 105$ $\quad$ $Z(C^{\pi^2}) = 102$ $\quad$ $Z(C^{\pi^3}) = 108$

$P^1 = 0.333$ $\quad$ $P^2 = 0.343$ $\quad$ $P^3 = 0.323$

(a) The candidate solutions and associated probabilities

$t = 1$

The element from $\pi^3 = (1,4,3,2)$ being followed:	3
Its location in the permutation $\pi^1 = (3,4,2,1)$:	(1,1)
The updated permutation $\pi^{1,1}$:	(2,4,3,1)
Updated circular matrix C:	$\begin{bmatrix} 30 & 22 & 13 & 40 \\ 23 & 7 & 36 & 29 \\ 30 & 22 & 7 & 32 \\ 36 & 32 & 23 & 17 \end{bmatrix}$
Maximum column sum $Z(C^{\pi^{1,1}})$:	119

$t = 2$

The element from $\pi^3 = (1,4,3,2)$ being followed:	1
Its location in the permutation $\pi^1 = (3,4,2,1)$:	(1,4)
The updated permutation $\pi^{1,2}$:	(1,4,2,3)
Updated circular matrix C:	$\begin{bmatrix} 22 & 13 & 40 & 30 \\ 23 & 7 & 36 & 29 \\ 22 & 7 & 32 & 30 \\ 23 & 17 & 36 & 32 \end{bmatrix}$
Maximum column sum $Z(C^{\pi^{1,2}})$:	144
$Z(C^{*\pi^1}) = \min\left(Z(C^{\pi^{1,1}}), Z(C^{\pi^{1,2}})\right)$ and	119 and
associated permutation π^1:	(2,4,3,1)

(b) Generation of Variations

Fig. 7.1 Illustrative example of CI solving the CBAP for a learning attempt

3. Using roulette wheel selection approach, assume that candidate 1 decides to follow candidate 3 and then generates two variations of the permutations $\Pi^{1,2} = \{\pi^{1,1}, \pi^{1,2}\}$ and associated maximum column sums $Z\left(C^{\pi^{1,1}}\right)$ and $Z\left(C^{\pi^{1,2}}\right)$ are calculated.
4. Further $Z\left(C^{\pi^1}\right) = \min\left(Z\left(C^{\pi^{1,1}}\right), Z\left(C^{\pi^{1,2}}\right)\right)$ and associated permutation $\pi^1 = \pi^{1,1}$ are identified.
5. In this way, candidates 2 and 3 also follow certain candidate in the cohort and find the $Z\left(C^{\pi^2}\right)$ and $Z\left(C^{\pi^3}\right)$ along with associated π^2 and π^3.

This process continues until convergence.

7.3.2 Numerical Experiments and Results

The CI algorithm discussed in Sect. 7.3 for solving the CBAP is coded in MATLAB 7.7.0 (R2008B). The simulations are run on a Windows platform using an Intel Core2 Quad CPU, 2.6 GHz processor speed and 4 GB memory capacity. The CI parameters such as number of candidates S and number of variations T are chosen to be 25 and 5, respectively. The problem size is determined by the order $n \times n$ of matrix C. In total, seventeen distinct cases with increasing problem size $n = 5$ to 13, 15, 20, 25, 30, 35, 40, 45, 50 are solved. For every case, 10 instances are generated and every instance is solved 20 times using the CI method. The CI saturation/convergence plot for problem instance $n = 30$ is presented in Fig. 7.2. This plot exhibits the self-adaptive learning behavior of every candidate in the cohort. Initially, the individual behavior/solution of every candidate in the cohort

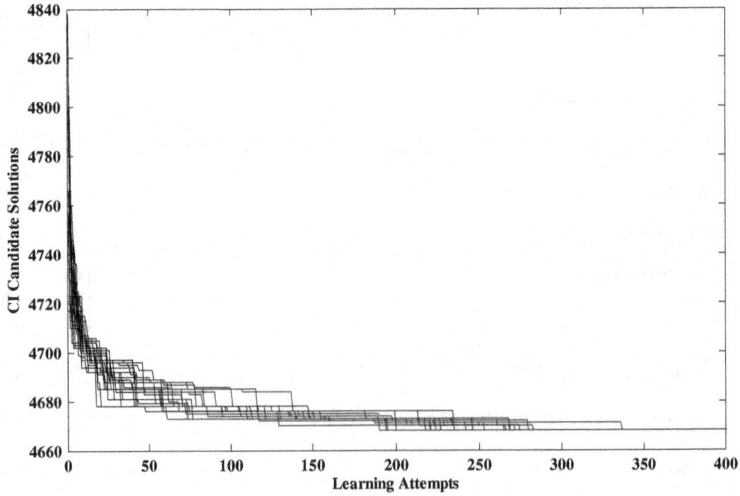

Fig. 7.2 Saturation/convergence of the cohort

can be easily distinguished. The behavior/solution here refers to the maximum column sum Z of the rotated matrix C^{π}. As every candidate adopts the qualities of other candidates to improve its own solution, the entire cohort gradually reaches a saturation stage and converges to an improved solution. It is important to mention here that the saturation associated restart procedure implemented in the original CI approach [2] which helped the candidates to explore further in the close neighborhood of their recently adopted qualities is not required.

In Table 7.1 we report our computational results obtained by solving the IP model given in Eqs. 7.1–7.6 for different values of the problem size n. For the solution quality, Table 7.1 shows the percentage gap between the best objective function values of solutions obtained using the LP relaxation of the model, CPLEX, the CI and MRSLS procedures. The percentage gap value between solution results of method X versus method Y is computed as $|Z_Y - Z_X| \times 100\,\%/Z_X$. These results are also summarized graphically in Fig. 7.3. First, as can be seen from columns 2 and 3 of Table 7.1, the LP relaxation of the model yields a tight lower bound that tends to improve as n is increased. This is a useful finding as it allows us to assess the performance of the CI method for large problem sizes. Indeed, as is evident from Fig. 7.3a, the times taken by CPLEX to solve the problem grow exponentially large as n increases. Unfortunately, we are able to report the CPU times for CPLEX only for n not exceeding 13. For n larger than 13, the times become prohibitively lengthy. That being said, a close examination of Table 7.1 reveals that the performance of CI method in solving CBAP is excellent both in terms of the percentage gap between the objective function values and the run times to solve the problem. For example, for $n = 13$, CPLEX takes close to 1073 s to reach an optimal solution whereas CI takes less than a second to produce a solution yielding an objective-value gap relative to CPLEX of less than 0.3 %. Also, the overall CPU time (refer to Fig. 7.3e) for CI is significantly less as compared to CPLEX. Furthermore, for comparatively smaller size cases, the solution obtained using CI method confirms with the CPLEX solution. An important observation from Table 7.1 and Fig. 7.3c, d is that similar to the percentage difference between the solution obtained using LP relaxation and CPLEX, the percentage difference between solution obtained using CI procedure and LP relaxation reduces gradually as the problem size increases. This demonstrates the noteworthy ability of CI in solving larger size problems with reasonable accuracy and also underscores its competitiveness with the CPLEX. Furthermore, the CI method could achieve the optimum solution for every case of the problem in reasonable number of function evaluations (FE). In addition, it is evident from Table 7.1 and Fig. 7.3f, g that the average number of FE is found to be increasing linearly while the standard deviation (SD) remains almost stable as the problem size increases. Since the search space increases with an increase in problem size, the number of characteristics a candidate learns from the other candidate being followed in a learning attempt do not change. This results into an increase in the number of learning attempts in order to improve their individual solution and eventually reach the saturation stage. Also, the SD presented in Table 7.1 demonstrates that the CI approach produces sufficiently robust solution for every case of the problem.

Table 7.1 Percentage gaps and CPU times of solutions obtained using the LP relaxation, CPLEX, and CI

Problem Size n	Average % gap LP versus CPLEX	CPLEX: average CPU time (s)	% gap LP versus MRSLS	% gap IP versus MRSLS	CI method					
					% gap CI versus MRSLS	Average % gap LP versus CI	Average % gap CPLEX versus CI	CI solution SD	CPU time (s)	FE
5	4.3165	0.27	4.5496	0	0	4.5496	0	0	0.0380	1620
6	3.3520	0.29	3.4857	0	0	3.4857	0	0	0.0824	3169
7	2.4426	0.41	2.5067	0	0	2.5067	0	0	0.1416	5070
8	1.7478	0.73	1.8423	0.0601	0.0598	1.7815	0	0.6597	0.2060	6988
9	1.3584	4.10	1.4842	0.1049	0.1046	1.3778	0	1.9202	0.3011	9136
10	0.9259	13.80	1.1638	0.2268	0.2139	0.9473	0.0120	3.2651	0.3597	10,750
11	0.8597	98.31	1.3708	0.4991	0.3912	0.9739	0.1057	2.9616	0.4433	12,190
12	0.7083	260.59	1.2766	0.5590	0.3422	0.9296	0.2146	3.0454	0.5370	13,601
13	0.5632	1072.93	1.2507	0.6804	0.3784	0.8675	0.2993	2.6952	0.6633	15,477
15	–	–	1.2879	–	0.4740	0.8040	–	2.8324	0.8314	17,827
20	–	–	0.9293	–	0.2772	0.6492	–	2.7274	1.6599	26,085
25	–	–	0.8442	–	0.2984	0.5431	–	2.6360	3.1387	35,435
30	–	–	0.6627	–	0.2087	0.4502	–	2.6683	5.0053	43,715
35	–	–	0.5657	–	0.1617	0.4031	–	2.5869	7.8426	51,145
40	–	–	0.5607	–	0.2177	0.3417	–	2.6962	11.9725	61,610
45	–	–	0.5086	–	0.1942	0.3130	–	2.4685	17.0199	71,835
50	–	–	0.4741	–	0.19482	0.2784	–	2.5459	23.9808	81,207

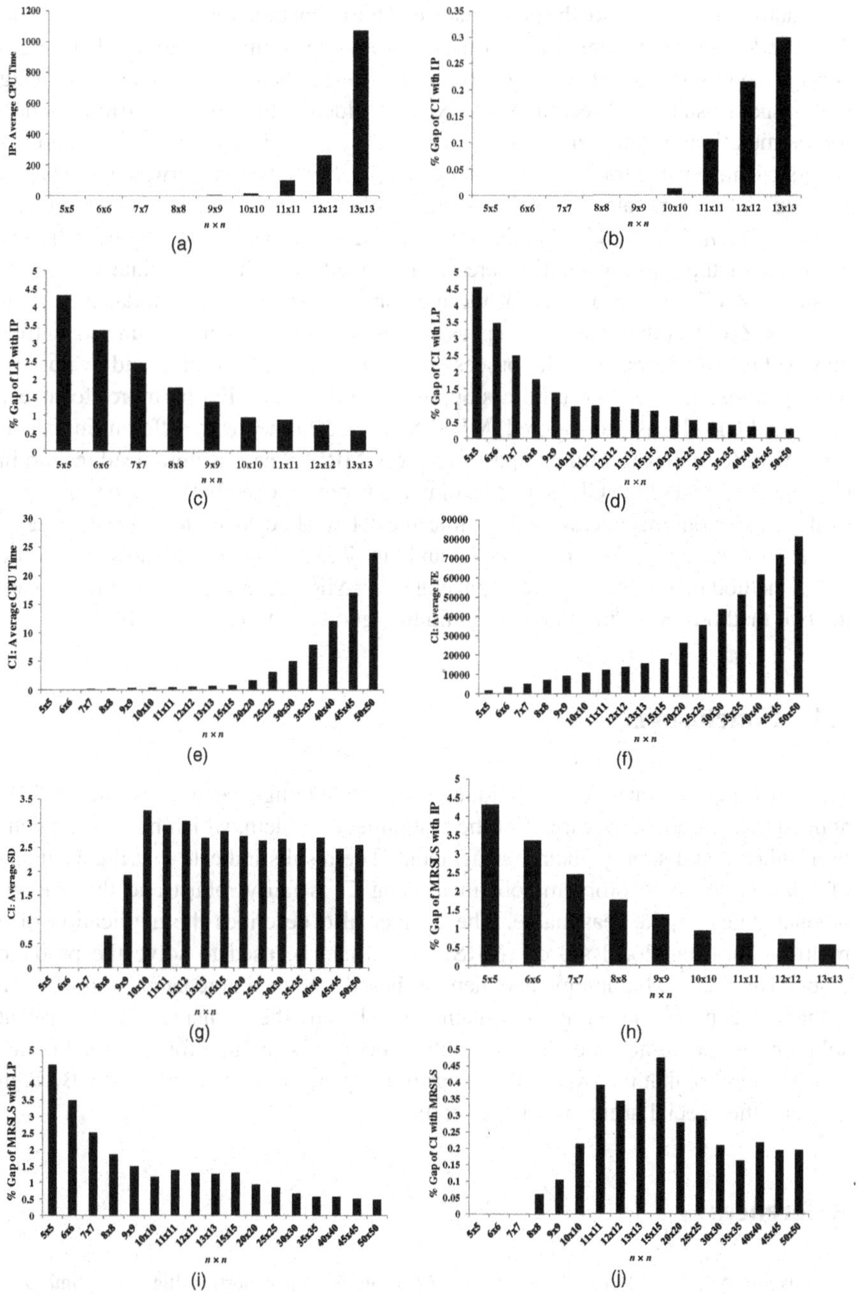

Fig. 7.3 Illustration of the CI, CPLEX, MRSLS and LP solution comparison

In addition, we compare the performance of the CI method for solving the CBAP to that of a local search technique that moves from one solution π_1 to another neighboring solution π_2 according to some prescribed rule. The following multi-random-start local search (MRSLS) is considered. In step 1, a starting solution (or permutation of the set $\{1, 2, \ldots, n\}$) $\pi_1 = (i_1, \ldots, i_k, \mathrm{i}, \mathrm{j}, \ldots, i_n)$ is randomly generated and a value for $Z(C^{\pi_1})$ is obtained. In step 2, we use a pairwise interchange approach to generate a neighboring solution π_2 which is given by $\pi_2 = (i_1, \ldots, i_k, \mathrm{j}, \mathrm{i}, \ldots, i_n)$; that is, the two elements $i$and j occupying adjacent positions in the current solution are interchanged. We then calculate the corresponding $Z(C^{\pi_2})$ value. In step 3, the incumbent best solution is updated to π_2 if $Z(C^{\pi_2}) < Z(C^{\pi_1})$; otherwise π_1 is kept as the best incumbent solution found so far (ties may be broken arbitrarily). The process is continued by performing and evaluating other pairwise interchange until a stopping criteria is met. Furthermore, for every individual CBAP case considered, MRSLS is run 50 times with different initialization. Also, for a meaningful comparison, every MRSLS case is initialized to start in the neighborhood of the CI's starting point and is run for exactly the same time equal to the corresponding average CPU time the CI method takes to solve that case. The results are summarized in Table 7.1 and Fig. 7.3c, h. The results show that, while the CI method in most cases has a slight edge over MRSLS in terms of optimality gap, the two methods perform quite well in finding good solutions to the CBAP.

7.4 Conclusions

The emerging optimization technique of cohort intelligence (CI) is successfully applied to solve the new variant of the assignment problem, which has applications in healthcare and supply chain management. The results indicate that the accuracy of solutions to these problems obtained using CI is fairly robust and the computational time is quite reasonable. The chapter also describes the application of a multi-random-start local search (MRSLS) that can be used to solve the problem cases. The MRSLS implemented here is based on the interchange argument, a valuable technique often used in sequencing, whereby the elements of two adjacent solutions are randomly interchanged in the process of searching for better solutions. Our findings are that the two methods perform equally well in solving the CBAP, in part due the special structure of the problem.

References

1. Kulkarni, A.J., Baki, M.F., Chaouch, B.A.: Application of the cohort-intelligence optimization method to three selected combinatorial optimization problems. Eur. J. Oper. Res. **250**(2), 427–447 (2016)
2. Kulkarni, A.J., Durugkar I.P., Kumar M.: Cohort intelligence: a self supervised learning behavior. In: Proceedings of IEEE International Conference on Systems, Man and Cybernetics, Manchester, UK, pp. 1396–1400, 13–16 Oct 2013

3. Kulkarni, A.J., Shabir, H.: Solving 0-1 Knapsack problem using cohort intelligence algorithm. Int. J. Mach. Learn. Cybernet. (2014). doi:10.1007/s13042-014-0272-y
4. Krishnasamy, G., Kulkarni, A.J., Paramesaran, R.: A hybrid approach for data clustering based on modified cohort intelligence and K-means. Expert Syst. Appl. **41**(13), 6009–6016 (2014)
5. Bartholdi, J.J., Orlin, J.B., Ratliff, H.D.: Cyclic scheduling via integer programs with circular ones. Oper. Res. **28**(5), 1074–1085 (1980)

Chapter 8
Solution to Sea Cargo Mix (SCM) Problem Using Cohort Intelligence Algorithm

Nomenclature

$\tilde{T}$ set of time periods $\{1, \ldots, t, \ldots, T\}$. Each period may represent one day, one week or one month, etc.

$\tilde{J}$ Set of ports of destinations for cargoes $\{1, \ldots, j, \ldots, J\}$

$\tilde{K}$ Set of all cargoes $\{1, \ldots, k, \ldots, K\}$ received in the planning horizon

η_k The period that cargo k will be received at the port of origin

τ_k The shipment due date for cargo k. Each cargo has its due date requested by shipper in its booking status

ξ_k The port of destination for cargo k. Cargo k will be received in period η_k and will be shipped to its destination port ξ_k on or before its due date τ_k

r_{kt} Per volume profit of cargo k which will be shipped in period t. It can be interpreted as the per volume net profit of cargo k, i.e., per volume revenue of cargo k minus its per volume delivery cost and inventory cost

E_t Total volume of available empty containers at the port of origin in period t

$V_{t,j}$ Total available volume capacity of shipment to port j in period t

$W_{t,j}$ Maximum allowable weight capacity of shipment to port j in period t

v_k Volume of cargo k

w_k Weight of cargo k

x_{kt} Binary variable, i.e., $x_{kt} = 1$, if cargo k is ready for shipment in period $t; x_{kt} = 0$, otherwise

The methodology of Cohort Intelligence (CI) [1–4] has been applied successfully applied solving combinatorial problems such as Knapsack problem, Traveling Salesman Problem and the new variant of the assignment problem (also referred to as Cyclic Bottleneck Problem (CBAP)). This chapter discusses CI solution to the Sea Cargo Mix (SCM) problem is originally proposed in [5]. The performance of CI solving the SCM is compared with the Integer Programming (IP) Solution as well as a multi-random-start local search (MRSLS) method. In addition the solution is compared with the Heuristic algorithm for MDMKP (HAM) and the Modified Heuristic algorithm for MDMKP (HAM) [5].

A.J. Kulkarni et al., *Cohort Intelligence: A Socio-inspired Optimization Method*, Intelligent Systems Reference Library 114, DOI 10.1007/978-3-319-44254-9_8

8.1 Sea Cargo Mix Problem

As mentioned before, the Sea Cargo Mix (SCM) problem is originally proposed in [5]. The decision problem consists of choosing a sea cargo shipping schedule of accepted freight bookings over a multi-period planning horizon. The goal is to maximize profit subject to constraints such as the limited available volume capacity, weight capacity and the number of available containers at the port of origin. The mathematical formulation of this problem, which can be viewed as a multi-dimension multiple knapsack problem (MDMKP), is discussed below.

$$Maximize\, Z = \sum_{1 \le k \le K} \sum_{\eta_k \le t \le \tau_k} v_k r_{kt} x_{kt} \tag{8.1}$$

Subject to

$$\sum_{k \in \tilde{K}_t} v_k x_{kt} \le E_t, \quad \forall t \in \tilde{T} \tag{8.2}$$

$$\sum_{k \in \tilde{K}_{tj}} v_k x_{kt} \le V_{tj}, \quad \forall t \in \tilde{T}, \ \forall j \in \tilde{J} \tag{8.3}$$

$$\sum_{k \in \tilde{K}_{tj}} w_k x_{kt} \le W_{tj}, \quad \forall t \in \tilde{T}, \ \forall j \in \tilde{J} \tag{8.4}$$

$$\sum_{\eta_k \le t \le \tau_k} x_{kt} \le 1, \quad \forall k \in \tilde{K} \tag{8.5}$$

$$x_{kt} \in \{0, 1\}, \quad \forall k \in \tilde{K}, \quad t\{\eta_k, \eta_k + 1, \ldots, \tau_k\} \tag{8.6}$$

where

$$\tilde{K}_t = \{k : k \in \tilde{K}, \quad \eta_k \le t \le \tau_k\}, \quad \forall t \in \tilde{T},$$
$$\tilde{K}_{tj} = \{k : k \in \tilde{K}, \quad \eta_k \le t \le \tau_k, \xi_k = j\}, \quad \forall t \in \tilde{T}, \ j \in \tilde{J}$$

The objective function (8.1) maximizes the total profit generated by all freight bookings accepted in the multi-period planning horizon T. Constraint (8.2) ensures that the demand for empty containers at the port of origin is less than or equal to the number of all available empty containers at the port of origin in each period. Constraint (8.3) ensures that the total volume of cargoes which will be carried to port j in period t is less than or equal to the total available volume capacity of shipment to port j in period t. Constraint (8.4) indicates that the total weight of cargoes which will be carried to port j in period t is less than or equal to the total available weight capacity of shipment to port j in period t. Constraint (8.5) stipulates that each cargo may be carried in a certain period on or before its due date or refused to be carried in the time horizon T. Constraint (8.6) states that each cargo is either accepted in its entirety or turned down.

There are J destination ports and T periods in the problem, and each cargo is either to be delivered within its due date or refused to be carried in the planning horizon. Thus, the total number of knapsacks is $T \times J$. Moreover, for each knapsack, there are three constraint sets, i.e., the set associated with the number of available empty containers, amount of available volume capacity and amount of available weight capacity.

8.2 Cohort Intelligence for Solving Sea Cargo Mix (SCM) Problem

In the context of CI algorithm presented in Chap. 2, the elements of cargo assignment set $C = k_t^{\xi_k}$ formed by assigning every cargo $k, \quad k \in \{1, 2, \ldots, K\}$ to a period $t \in \{1, 2, \ldots, T\}$ being shipped to its port of destination ξ_k are considered as characteristics/attributes/qualities of the cohort candidate. The port of destination ξ_k for every cargo $k \in \{1, 2, \ldots, K\}$ is selected based on the condition below.

$$\begin{array}{lll} \xi_k = j, & if\ [K/J] \times (j-1) < k \leq [K/J] \times j, & for\ j = 1, 2, \ldots, J-1 \\ \xi_k = J, & if\ [K/J] \times (j-1) < k \leq K, & \text{otherwise} \end{array} \tag{8.7}$$

The CI algorithm begins with the initialization of number of cohort candidates S, number of variations Y the cargo assignment set $\mathrm{C^s}$ of every candidate s, $(\mathrm{s} = 1, \ldots, \mathrm{S})$, the convergence parameter ε and maximum number of allowable learning attempts L_{max}.

In the cohort, every candidate $s, (s = 1, \ldots, S)$ randomly assigns every cargo $c_k, \quad k \in \{1, 2, \ldots, K\}$ to a period $t \in \{1, 2, \ldots, T\}$ to be shipped to destination ξ_k and forms a cargo assignment set (behavior) $C^s = k_t^{s, \xi_k}$ and associated per volume profit are calculated as $R^s = \sum_{k=1}^{K} \sum_{t=1}^{1} r_{k,t}^s$.

Step 1. **(Constraint Handling)** As a maximization problem, the probability associated with per volume profit of cargo R^s is calculated as follows:

$$p_R^s = \frac{R^s}{\sum_{s=1}^{S} R^s}, \quad (s = 1, \ldots, S) \tag{8.8}$$

There are constraints involved such as:

1. demand of empty containers $\sum_k v_{k,t}^s$ at the port of origin should be less than or equal to the number of all available empty containers E_t at the port of origin in each period $t \in \{1, 2, \ldots, T\}$

2. total volume of cargoes $\sum_k v^s_{k,j}$ which will be carried to port $j \in \{1, 2, \ldots, J\}$ in period $t \in \{1, 2, \ldots, T\}$ is less than or equal to the total available volume capacity $V_{t,j}$, and
3. total weight of cargoes $\sum_k v^s_{k,j}$ which will be carried to port $j \in \{1, 2, \ldots, J\}$ in period $t \in \{1, 2, \ldots, T\}$ is less than or equal to the corresponding total available weight capacity $W_{t,j}$.

Kulkarni and Shabir [3] propose a modified approach to the CI method for solving knapsack problems. This approach makes use of probability distributions for handling constraints. This approach is also adopted here. For every constraint type as described in 1, 2 and 3 above a probability distribution is developed (refer to Fig. 8.1) and the probability is calculated based on the following rules:

1. If $0 \leq \sum_k v^s_{k,t} \leq E_t$, $\forall t$, then based on the probability distribution presented in Fig. 8.1a $p^s_{E_t} = slope_{1,Et} \times \left(\sum_k v^s_{k,t} - E_t\right)$, else $p^s_{E_t} = slope_{1,Et} \times (0.001\ \%E_t)$.
2. If $0 \leq \sum_k v^s_{k,j} \leq V_{t,j}$, $\forall t, \forall j$, then based on the probability distribution presented in Fig. 8.1b $p^s_{V_{t,j}} = slope_{1,V_{t,j}} \times \left(\sum_k v^s_{k,j} - V_{t,j}\right)$, else $p^s_{V_{t,j}} = slope_{1,V_{tj}} \times \left(0.001\ \%V_{t,j}\right)$.
3. If $0 \leq \sum_k w^s_{k,j} \leq W_{t,j}$, $\forall t, \forall j$, then based on the probability distribution presented in Fig. 8.1c $p^s_{W_{t,j}} = slope_{1,W_{t,j}} \times \left(\sum_k w^s_{k,j} - W_{t,j}\right)$, else $p^s_{W_{tj}} = slope_{1,W_{tj}} \times \left(0.001\ \%W_{t,j}\right)$.

As represented in Fig. 8.1, the $slope_{1,Et}$, $slope_{1,V_{t,j}}$ and $slope_{1,W_{t,j}}$ represent the slope of lines going through points $((0,1),(E_t,0))$, $((0,1),(V_{tj},0))$ and $((0,1),(W_{tj},0))$, respectively. The overall (total) probability of selecting candidates to follow candidate s, $(s = 1, \ldots, S)$ is calculated as follows:

$$p^s = \left(p^s_R + \sum\nolimits_t p^s_{E_t} + \sum\nolimits_t \sum\nolimits_j p^s_{V_{t,j}} + \sum\nolimits_t \sum\nolimits_j p^s_{W_{t,j}}\right) \tag{8.9}$$

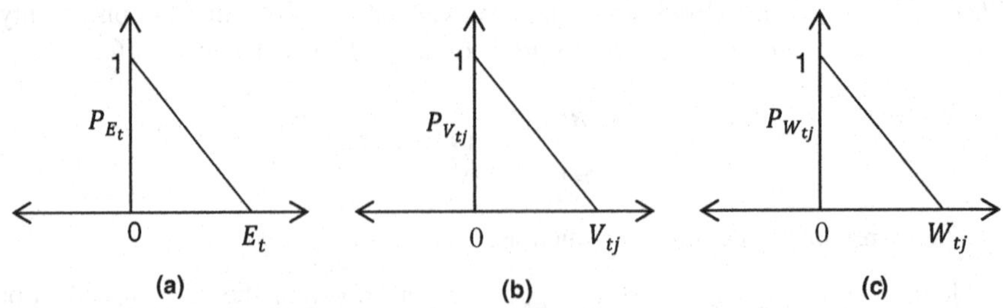

Fig. 8.1 Probability distributions for constraint handling

It is clear from the above rules for probability calculation that the candidate's behavior/solution/cargo assignment with better objective and constraint values closer to the boundaries will have higher probability of being followed.

Step 2. Every candidate generates Y new variations of the cargo assignment using two steps, which we refer to as 'learning from others' and 'introspection', as follows:

1. **Learning from others**: Every candidate $s, \quad (s = 1, \ldots, S)$ using roulette wheel approach [1–4] selects a candidate $\widehat{s} \in (1, \ldots, S)$ (not known in advance) in the cohort to follow, i.e. it incorporates an element from within $c^{\widehat{s}}$ into its existing cargo assignment c^s. More specifically, a quality from within $c^{\widehat{s}}$ is selected randomly. Then the selected element is identified in c^s along with its location. It then swaps its position with the element at the location in c^s corresponding to its current location in $c^{\widehat{s}}$. This way every candidate $s, \quad (s = 1, \ldots, S)$ generates $Y/2$ cargo assignments.
2. **Introspection**: In addition, every candidate $s, \quad (s = 1, \ldots, S)$ randomly selects an element from within its one of the periods $t, \quad (t = 1, \ldots, T)$ and relocates it to another period. This way every candidate $s, \quad (s = 1, \ldots, S)$ generates further $Y/2$ cargo assignments.

This way every candidate forms a total of Y new variations $\mathrm{C}^{s,Y} = \{\mathrm{c}^{s,1}, \ldots, \mathrm{c}^{s,y}, \ldots, \mathrm{c}^{s,Y}\}$ and computes associated per volume profit and constraint functions.

Step 3. As discussed in Step 1, every candidate $s, \quad (s = 1, \ldots, S)$ calculates its corresponding probability vector $P^{s,Y} = \{p^{s,1}, \ldots, p^{s,y}, \ldots, p^{s,Y}\}$. Furthermore, based on the feasibility-based rules shown below, the candidate accepts or rejects the solution associated with the maximum total probability value, i.e. $max\{p^{s,1}, \ldots, p^{s,y}, \ldots, p^{s,Y}\}$

The feasibility-based rules are as follows:
Accept the current behavior/solution if

1. The cargo assignment in the previous learning attempt is feasible and current behavior/cargo assignment is also feasible with improved per volume profit
2. The cargo assignment in the previous learning attempt is infeasible and the current behavior/cargo assignment is feasible
3. The cargo assignment in the previous learning attempt is infeasible and the current behavior/cargo assignment is also infeasible with the maximum total probability value improved;

Otherwise, reject the current behavior/solution and retain the previous one if

1. The cargo assignment in the previous learning attempt is feasible and current cargo assignment is infeasible
2. The cargo assignment in the previous learning attempt is feasible and the current cargo assignment is also feasible with worse per volume profit
3. The cargo assignment in the previous learning attempt as well as current learning attempt are infeasible and the total probability value is lesser than the previous learning attempt.

After the completion of step 3, a cohort with S updated cargo assignments $\{c^1, \ldots, c^s, \ldots, c^S\}$ is now available.

Step 4. If either of the two criteria listed below is valid, accept the best possible cargo assignment from within the available $\{c^1, \ldots, c^s, \ldots, c^S\}$ in the cohort as the final solution c^* and stop, else continue to Step 1

(a) If maximum number of learning attempts exceeded or
(b) The cohort is saturated, i.e. if cohort candidates saturate to the same cargo assignment for any other number of successive learning attempts.

8.3 Numerical Experiments and Results

The following notation is used to describe the results of our numerical experiments:

N_v	Number of decision variable in the problem
N_c	Number of constraints in the problem
N_{in}	Number of tested instances
U	Upper bound
I	Integer programming solution (branch-and-bound method)
L	LP relaxation
H	Heuristic algorithm for MDMKP (HAM) (refer to [5])
M	Modified heuristic algorithm for MDMKP (MHA) (refer to [5])
CI	Cohort intelligence (CI) method
MRSLS	Multi-Random-Start Local Search
g_{XZ}	Average percentage gap between the best objective value of the solutions obtained using methods X and Z
$g_{\widetilde{XZ}}$	Average percentage gap between the average objective value of the solutions obtained using methods X and Z
$g_{\widehat{XZ}}$	Worst percentage gap between the worst objective value of the solutions obtained using methods X and Z

(continued)

(continued)

t_X	Average computational time (in seconds) of algorithm X
$S_{t_{CI}}$	Standard deviation of CPU time for CI method
S_{XZ}	Standard deviation of percentage gap between objective value of the solutions obtained using methods X and Z

The CI approach for solving the SCM Problem discussed in Sect. 8.1 is coded in MATLAB 7.7.0 (R2008B). The simulations are run on a Windows platform with an Intel Core2 Quad CPU, 2.6 GHz processor speed and 4 GB memory capacity. For this model, we solve 18 distinct cases. These cases, which are originally proposed in [5], are presented in Tables 8.1, 8.2 and 8.3. For every case, 10 instances are generated and every instance is solved 10 times using the CI method. The instances are generated as suggested in [5]. The per volume profit $r_{k,t}$ for cargo k, $(k = 1,\ldots,K)$ shipped in period t, $(t = 1,\ldots,T)$ are uniformly generated in the interval $[0.01, 1.01]$. The volume v_k and weight w_k of every cargo c_k, $(k = 1,\ldots,K)$ are uniformly generated from the interval $[100, 200]$. The number of all available empty containers E_t at the port of origin in each period $t \in \{1,2,\ldots,T\}$ are uniformly generated from the interval $[100 \times (K/T), 200 \times (K/T)]$, and the total volume $V_{t,j}$ and weight $W_{t,j}$ of cargoes which are carried to port $j \in \{1,2,\ldots,J\}$ in period $t \in \{1,2,\ldots,T\}$ are uniformly generated from the interval $[100 \times (K/T), 200 \times (K/T)]$.

The CI parameters such as number of candidates S and number of variations Y are chosen to be 3 and 15, respectively. The CI saturation/convergence plot for one problem instance given by $(T, J, K) = (4, 13, 5479)$ is presented in Fig. 8.2. The plot exhibits the self-adaptive learning behavior of every candidate in the cohort. Initially, the distinct behavior/solution of every individual candidate in the cohort can be easily distinguished. The behavior/solution here refers to the total profit generated by all freight bookings accepted in the multi-period planning horizon T. As each candidate adopts the qualities of other candidates to improve its own behavior/solution, the behavior of the entire cohort saturates/converges to an improved solution.

The best and average CI solution for the objective function value for every case is compared with the associated upper bound (UB) solution achieved by solving the LP relaxation of the problem, and the integer programming (IP) solution. In addition, the solution is compared to the solution of the LP relaxation, and the problem-specific heuristic algorithm for MDMKP (HAM) and the modified heuristic algorithm for MDMKP (MHA) developed in [5]. The numerical results are presented in Tables 8.1, 8.2 and 8.3 along with the graphical illustration in Fig. 8.3. It is important to mention here that IP is not able to solve large-scale SCM problems.

Table 8.1 Results for small scale test problems

T, J, K	N_v	N_c	N_{in}	IP	LP (L)	HAM			MHA		
				CPU time (s) t_I	CPU time (s) t_L	g_{IH}	g_{LH}	CPU time (s) t_I	CPU time (s) t_L	g_{IH}	g_{LH}
3, 5, 41	123	74	10	0.106	0.028	1.59	2.88	0.001	1.55	2.71	0.015
3, 6, 47	141	86	10	0.274	0.047	1.26	2.67	0.002	1.03	2.42	0.023
4, 3, 64	256	92	10	0.480	0.183	0.87	2.32	0.011	0.67	1.53	0.056
2, 4, 132	264	150	10	0.148	0.391	0.68	2.03	0.016	0.51	1.18	0.093
3, 3, 91	273	112	10	0.257	0.289	0.93	1.98	0.008	0.78	1.79	0.046
2, 3, 143	286	157	10	0.096	0.485	0.46	1.03	0.014	0.28	0.68	0.078

CI Performance

Best sol % gap	Avg sol % gap	Worst sol % gap	Best sol % gap	Avg sol % gap	Worst Sol % gap	CPU time (s)	SD (CPU time)	SD	SD
0.5375	1.3188	2.3818	0.0237	0.2452	1.8782	0.070	0.036	0.691	0.687
0.3902	1.0102	1.6991	0.0709	0.6931	1.3842	0.093	0.027	0.456	0.454
0.5466	0.9958	1.5231	0.0616	0.5131	1.0430	0.103	0.025	0.351	0.349
0.4813	1.0906	1.6339	0.3960	1.0059	1.5496	0.098	0.041	0.408	0.408
0.3758	0.9250	1.6017	0.1720	0.8474	1.4006	0.130	0.046	0.463	0.462
0.1827	0.8045	1.3027	0.1424	0.7644	1.2629	0.094	0.037	0.379	0.378

It is evident from the results in Tables 8.1, 8.2, 8.3 and the plots given in Fig. 8.3a, d that, for small scale SCM problems, the CI method produces a solution that is fairly close to the IP and UB solution. The gap gradually increases as the problem size grows; however, observe that the worst gap between the best CI solution and corresponding IP (g_{ICI}) and UB solution (g_{ICI}) is within 1.0459 % of the reported IP solution and 4.0405 % of the reported UB solution, respectively. Similarly, the worst gap between the average CI solution and corresponding IP $(g_{\widetilde{ICI}})$ and UB solution $(g_{\widetilde{ICI}})$ is within 2.2682 % of the reported IP solution and 5.5827 % of the reported UB solution, respectively. Also, the percent gap between the worst CI solution and corresponding IP solution $(g_{\widehat{ICI}})$ is within 3.0198 % of the reported IP solution. The corresponding UB solution $(g_{\widehat{ICI}})$ is within 7.1465 % of the reported UB solution.

Furthermore, as shown in Tables 8.1, 8.2, 8.3 and Fig. 8.3i, j, even though the standard deviation (SD) of the percent gap between the CI solution and the corresponding IP (S_{ICI}) and UB solution (S_{ICI}) increases with the problem size, the

Table 8.2 Results for medium scale test problems

T, J, K	N_v	N_c	N_{in}	IP	LP (L)	HAM			MHA		
				CPU time (s) t_I	CPU time (s) t_L	g_{IH}	g_{LH}	CPU time (s) t_I	CPU time (s) t_L	g_{IH}	g_{LH}
3, 4, 900	2700	927	10	0.722	419.1	3.92	1.12	1.72	6.17	3, 4, 900	2700
4, 8, 965	3860	1033	10	1.833	1264.3	2.66	2.58	1.16	14.37	4, 8, 965	3860
4, 25, 1000	4000	1204	10	1.240	2012.5	2.05	8.12	0.86	49.66	4, 25, 1000	4000
2, 3, 2871	5742	2885	10	1.227	5872.2	1.35	5.11	0.75	28.29	2, 3, 2871	5742
2, 3, 3876	7752	3890	10	1.887	199966	0.56	5.97	0.32	55.41	2, 3, 3876	7752
5, 37, 1954	9770	2329	10	4.151	12306.1	1.83	57.67	1.26	321.02	5, 37, 1954	9770

CI Performance

Best sol % gap	Avg sol % gap	Worst sol % gap	Best sol % gap	Avg sol % gap	Worst Sol % gap	CPU time (s)	SD (CPU time)	SD	SD
1.0516	2.2738	3.0253	1.0459	2.2682	3.0198	0.734	0.271	0.650	0.650
0.9356	1.3321	1.8138	0.9284	1.3250	1.8067	1.382	0.609	0.299	0.299
0.0859	0.8452	1.6848	0.0794	0.8387	1.6785	0.185	0.225	0.611	0.611
1.0009	1.7398	2.2239	0.9968	1.7357	2.2199	2.931	0.798	0.380	0.380
0.5979	1.1463	1.5484	0.5944	1.1428	1.5449	2.488	0.586	0.303	0.303
0.4617	1.4282	2.5431	0.4576	1.4241	2.5390	1.305	0.425	0.917	0.917

worst SD is 0.917. Moreover, Table 8.1, 8.2, 8.3 and Fig. 8.3f also show that the SD ($S_{t_{CI}}$) of CPU time for solving small- and medium-scale problems is within 0.046 and 0.708, respectively. For large-scale problems it is within 4.940. This is because the search space increases with an increase in problem size.

For every candidate the number of characteristics to be learnt in a learning attempt from the candidate that is being followed does not change. This results into different number of learning attempts to improve their individual behavior/solution and to eventually reach the saturation/convergence state. However, it is important to mention here that the overall SD obtained by solving the entire problem set is quite reasonable which lends support to the robustness of the algorithm.

Also, the percent gap between the worst CI solution and corresponding IP solution $\left(g_{ICI}^{\wedge}\right)$ is within 3.0198 % of the reported IP solution. The corresponding UB solution $\left(g_{UCI}^{\wedge}\right)$ is within 7.1465 % of the reported UB solution. This

Table 8.3 Results for large scale test problems

T, J, K	N_v	N_c	N_{in}	HAM	MHA	CI Performance					
				CPU time (s) t_H	CPU time (s) t_M	Best sol % gap g_{UCI}	Avg sol % gap $\tilde{g}_{UCI}$	Worst sol % gap $\hat{g}_{UCI}$	CPU time (s) t_{CI}	SD (CPU time)	SD S_{UCI}
9, 47, 1521	13689	2376	10	80.5	447.1	3.3822	4.6675	5.8294	6.7303	0.344	0.707
3, 4, 6576	19728	6603	10	53.2	293.1	3.8078	5.2705	6.6293	61.6805	4.940	0.726
4, 5, 5286	21144	5330	10	54.5	277.7	3.6931	5.1307	6.5576	35.6223	3.819	0.835
4, 13, 5479	21916	5587	10	125.9	785.9	3.8556	5.5827	7.1253	27.2433	2.885	0.898
5, 8, 4954	24770	5039	10	89.9	454.2	4.0504	5.5827	7.0751	41.9530	3.514	0.766
8, 26, 3249	25992	3673	10	178.3	982.9	3.6647	5.2144	7.1465	28.5853	1.212	0.892

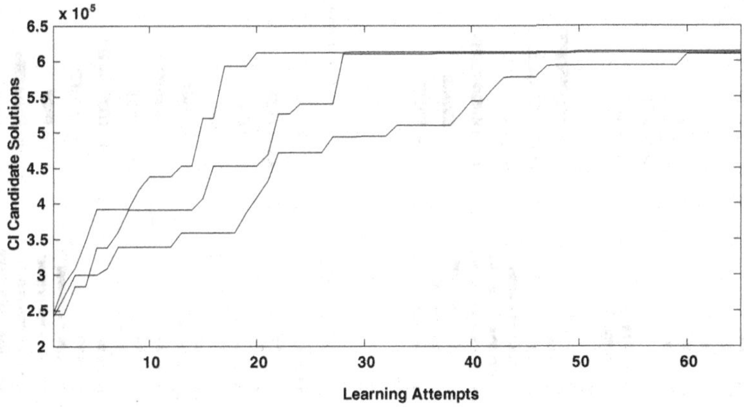

Fig. 8.2 Saturation/convergence of the cohort for instance of the SCM problem

demonstrates that, even though the magnitude of S_{ICI}, S_{UCI} and $S_{t_{CI}}$ increases with increase in problem size, CI is able to produce solutions with reasonable accuracy for every case of the problem. In addition, the CI method achieves the optimum solution for medium- and large-scale problems in significantly less CPU time (refer to Fig. 8.3h). This demonstrates the ability of CI in solving large problems efficiently and highlights its competitiveness with the IP approach as well as the heuristics HAM and MHA discussed in [5].

In addition to the above, CI's performance is also compared to the performance of a multi-random-start local search (MRSLS) that is used to solve the Sea Cargo Mix problem. The proposed MRSLS follows a similar pairwise interchange approach that we use for the CBAP discussed in Chap. 7. For each of the problem instances suggested in [5], a solution is first constructed. Then a pairwise interchange approach is used in every successive learning attempt where two time periods are selected randomly. Next a set of containers associated with each period is randomly chosen and then the positions of these two sets are interchanged (swapped). The MRSLS for every individual case of the SCM problem is run 50 times with different initializations. Also, for a meaningful comparison, every MRSLS case is initialized to start in the neighborhood of the CI's starting point and is run for exactly the same time equal to the corresponding average CPU time the CI method takes to solve that case. The acceptance of the resulting solution in every learning attempt depends on following feasibility-based rules (see [6] for a detailed discussion): (1) if the existing solution is infeasible and the resulting solution has improved constraint violation, then the solution is accepted, (2) If the existing

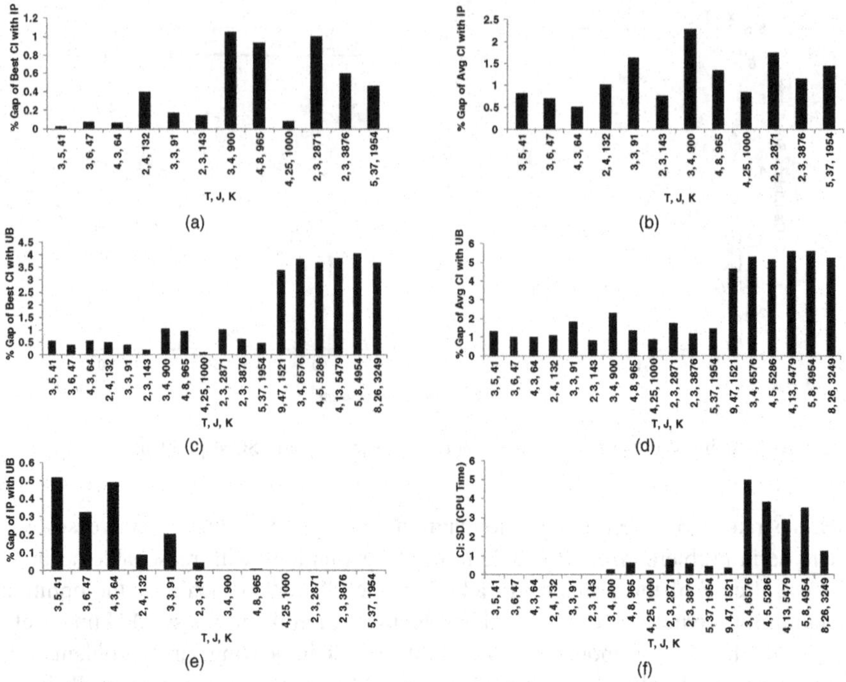

Fig. 8.3 Illustration of CI, IP, MRSLS and UB solution comparison

solution is infeasible and the resulting solution is feasible, then the solution is accepted, (3) if the existing solution is feasible and the resulting solution is also feasible yielding an improved objective function value Z, then the solution is accepted. If any of these conditions are not satisfied then the existing solution is retained and the resulting solution is discarded.

It is important to mention here that of the 50 MRSLS runs related to the SCM problems under study, only a few of the solutions obtained are feasible. Most of solutions are outside the feasible region. This is because for every MRSLS run a starting solution is randomly chosen and this solution can be infeasible. Furthermore, the MRSLS may not be able to discover a feasible solution during the entire run. Therefore, only the best of the feasible solutions are considered for meaningful comparison with the CI approach. From Tables 8.4, 8.5 and 8.6 as well as Fig. 8.3a, k it is clear that the rate of increase of the percentage gap between the solution obtained using MRSLS and that obtained using CPLEX is significantly more when compared to the rate of percentage gap increase between CPLEX and

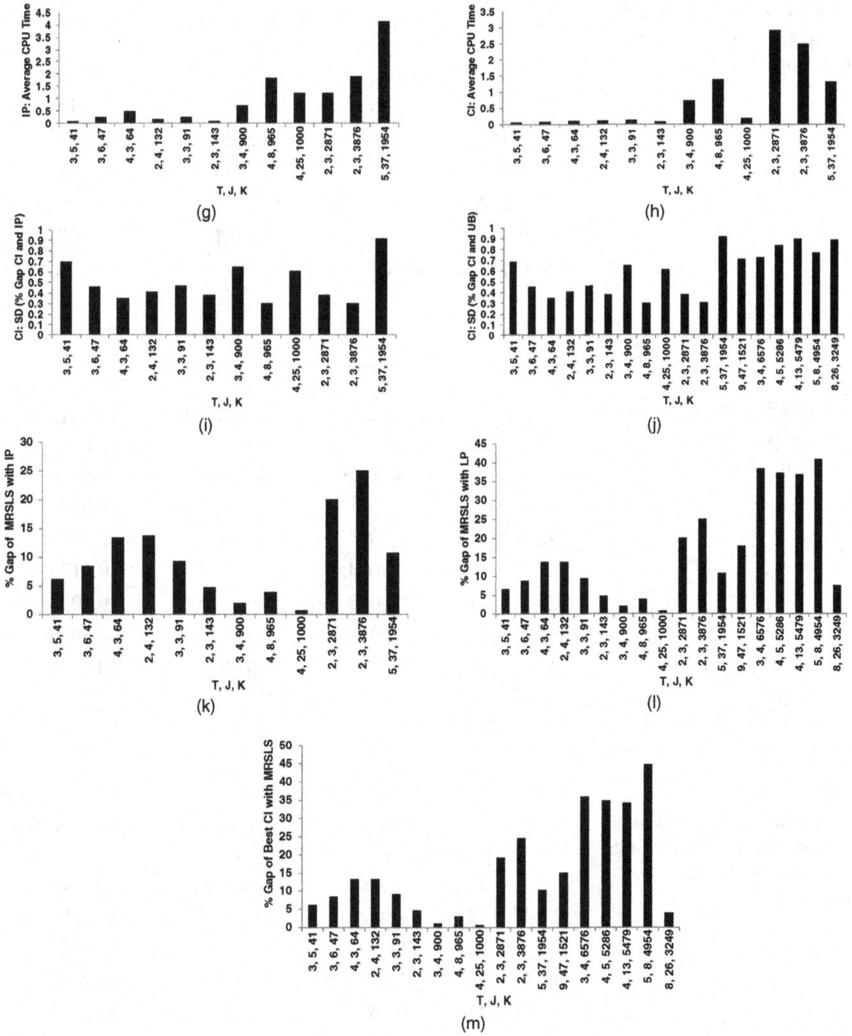

Fig. 8.3 (continued)

CI. In addition, the percentage gap between the solution obtained using MRSLS and LP relaxation for each case is also considerably larger as compared to that of CI versus LP relaxation. In short, for the Sea Cargo Mix problem, CI achieves better performance against the MRSLS implemented for this model, especially when the problem size is large.

Table 8.4 MRSLS results comparison for small scale test problems

T, J, K	Best sol % gap $g_{MRSLSLP}$	Best sol % gap $g_{MRSLSLP}$	Best sol % gap $g_{MRSLSLP}$
3, 5, 41	6.6548	6.1737	6.1516
3, 6, 47	8.8023	8.5057	8.4391
4, 3, 64	13.8332	13.4053	13.3487
2, 4, 132	13.7926	13.7174	13.3799
3, 3, 91	9.5009	9.3181	9.1595
2, 3, 143	4.8001	4.7614	4.6253

Table 8.5 MRSLS results comparison results for medium scale test problems

T, J, K	Best sol % gap $g_{MRSLSLP}$	Best sol % gap $g_{MRSLSIP}$	Best sol % gap $g_{MRSLSIP}$
3, 4, 900	2.0691	2.0635	1.0303
4, 8, 965	3.9044	3.8974	3.0055
4, 25, 1000	0.7014	0.6951	0.6178
2, 3, 2871	20.0099	20.0065	19.2088
2, 3, 3876	25.0458	25.0432	24.5916
5, 37, 1954	10.7262	10.7223	10.3139

Table 8.6 MRSLS results comparison results for large scale test problems

T, J, K	Best sol % gap $g_{MRSLSLP}$	Best sol % gap $g_{MRSLSIP}$	Best sol % gap $g_{MRSLSIP}$
9, 47, 1521	17.9624	–	15.1393
3, 4, 6576	38.3933	–	35.9608
4, 5, 5286	37.1910	–	34.7788
4, 13, 5479	36.7655	–	34.2314
5, 8, 4954	40.8499	–	44.6732
8, 26, 3249	7.5249	–	4.0031

8.4 Conclusions

The emerging optimization technique of cohort intelligence (CI) is successfully applied to solve a complex combinatorial problem such as the sea cargo mix problem. For the problem a specific CI algorithm is developed. The results indicate that the accuracy of solutions to these problems obtained using CI is fairly robust and the computational time is quite reasonable. The chapter also describes the application of a MRSLS that can be used to solve several cases of the problem.

The MRSLS implemented here is based on the interchange argument, a valuable technique often used in sequencing, whereby the elements of two adjacent solutions are randomly interchanged in the process of searching for better solutions. Our findings are that the performance of the CI is clearly superior to that of IP, HAM and MHA as well as the MRSLS for most of the problem instances that have been solved.

In agreement with the no-free-lunch theorem [7], any algorithm may not be directly applicable to solve all the problem types unless it can be enhanced by incorporating some useful techniques or heuristics. The CI method may also benefit from certain performance-enhancing techniques when it is applied to different classes of problems. A mechanism to solve multi-objective problems is currently being developed, which can prove helpful in transforming the model's constraints into objectives/criteria (see [7] for new development in this area). This can help reduce the dependency on the quality of the candidates' initial guess.

References

1. Kulkarni, A.J., Durugkar I.P., Kumar M.: Cohort intelligence: a self supervised learning behavior. In: Proceedings of IEEE International Conference on Systems, Man and Cybernetics, Manchester, UK, pp. 1396–1400, 13–16 Oct 2013
2. Kulkarni, A.J., Baki, M.F., Chaouch, B.A.: Application of the cohort-intelligence optimization method to three selected combinatorial optimization problems. Eur. J. Oper. Res. (2015)
3. Kulkarni, A.J., Shabir, H.: Solving 0-1 Knapsack Problem using cohort intelligence algorithm. Int. J. Mach. Learn. Cybern. (2014). doi:10.1007/s13042-014-0272-y
4. Krishnasamy, G., Kulkarni, A.J., Paramesaran, R.: A hybrid approach for data clustering based on modified cohort intelligence and K-means. Expert Syst. Appl. **41**(13), 6009–6016 (2014)
5. Ang, J.S.K., Cao, C., Ye, H.Q.: Model and algorithms for multi-period sea cargo mix problem. Eur. J. Oper. Res. **180**(3), 1381–1393 (2007)
6. Deb, K.: An efficient constraint handling method for genetic algorithms. Comput. Methods Appl. Mech. Eng. **186**(2–4), 311–338 (2000)
7. Patankar, N.S., Kulkarni, A.J., Tai, K., Ghate, T.D., Parvate, A.R.: Multi-criteria probability collectives. Int. J. Bio-Inspired Comput. **6**(6), 369–383 (2014)

Chapter 9
Solution to the Selection of Cross-Border Shippers (SCBS) Problem

Nomenclature

T	Number of periods in the planning horizon
I	Number of containers
J	Number of shippers
K	Number of different types of goods
i	Containers, $i = 1, 2, \ldots, I$
j	Shippers, $j = 1, 2, \ldots, J$
k	Type of goods, $k = 1, 2, \ldots, K$
Δ_k	Set of containers of type k
α_{jk}	A binary parameter = 1, if shipper j can handle the containers of type k = 0, otherwise
a_j	Fixed cost of choosing shipper j
b_{ij}	Variable cost of shipping container i through shipper j. If a shipper cannot handle the kind of goods in a container, then the variable cost is set to a high value
c_j	Maximum capacity of shipper j
e_i	Volume of container i
p_{ij}	Expected time of processing container i through shipper j
D_i	Due date of container i
F	Fund available
w_1	Weight assigned to the goal of fund constraint
w_{2i}	Weight assigned to the goal of due date for container i
w_{31}	Penalty for exceeding the limit of non-compliant shippers
w_{32}	Reward for using fewer than allowable number of non-compliant shippers
θ_j	A binary parameter = 1, if shipper j complies with cross-border regulations = 0, otherwise
θ^0	Maximum allowable number of non-compliant shippers
$\hat{c}_{jt}$	Maximum capacity of shipper j in period t
$\hat{F}_t$	Fund available in period t
t	Index for periods, $t = 1, 2, \ldots, T$
N_v	Number of decision variables in the problem

A.J. Kulkarni et al., *Cohort Intelligence: A Socio-inspired Optimization Method*, Intelligent Systems Reference Library 114, DOI 10.1007/978-3-319-44254-9_9

N_c	Number of constraints in the problem
N_{in}	Number of tested instances
d_1^- and d_1^+	Deviational variables associated with the fund constraint in the one-period setting
d_{1t}^- and d_{1t}^+	Deviational variables associated with the fund constraint in period t
d_{2i}^- and d_{2i}^+	Deviational variables associated with every container i, $i = 1, 2, \ldots, I$
d_3^- and d_3^+	Deviational variables associated with the selection of the non-compliant shippers θ^0

In this chapter, we demonstrate the ability of Cohort Intelligence (CI) methodology to solve problem of optimal selection of cross-border shippers and cargo assignments [1]. The problem includes various constraints related to due dates, processing times, fund availability, and shippers' compliance. We formulate and solve the multi-period instance of this problem as well. The performance of the CI method is compared to that of Integer Programming (IP) solution obtained using CPLEX and to specifically developed multi-random-start local search (MRSLS) method.

9.1 Selection of Cross-Border Shippers (SCBS) Problem

Cross-border shippers are major players in international trade and transportation [1, 2]. With the ever-changing standards of international compliance, international shippers of imported and exported goods must comply with an increasing number of regulatory constraints. The selection of shippers with cross-border compliance/non-compliance emerged as an important problem after the North American Free Trade Agreement (NAFTA) [3] became functional in 1994 which considerably increased cargo traffic between Canada, the United States and Mexico. Selecting compliant cross-border shippers helps avoid frustrating shipment delays at border check points and also results in transportation cost savings. In this section, we examine the problem of a company that must meet a number of goals by selecting shippers for the purpose of transporting containerized cargo across borders. The company must rely on shippers that can be either compliant or non-compliant. On the one hand, a compliant shipper is more costly to use; however, it allows shorter delivery times as it can facilitate the smooth transit of cargo through the border. On the other hand, a non-compliant shipper, while cheaper to use, may take longer delivery times of the cargo to the customer's destination as it may experience inspection slowdowns at the border. The elements involved in selection of a cross-border shipper problem include the total cargo volume to be transported to customers, the total funds available over the planning period, the ability of shippers to handle special types of goods, anticipated delivery due dates, processing times for the particular good through the shipper, the type of shipper to use, etc. The mathematical formulations of the single- and multi-period problems are discussed below.

9.1.1 Single Period Model

Defining the decision variables as

$$x_{ij} = \begin{cases} 1 & \text{if container } i \text{ is shipped through shipper } j \\ 0 & \text{otherwise} \end{cases}$$

$$y_j = \begin{cases} 1 & \text{if shipper j is chosen} \\ 0 & \text{otherwise} \end{cases}$$

leads to the following formulation:

$$Min\, w_1 d_1^+ + \sum_i w_{2i} d_{2i}^+ + w_{31} d_3^+ - w_{32} d_3^- \quad (9.1)$$

$$\sum_i e_i x_{ij} \le c_j \quad \forall j \quad (9.2)$$

$$x_{ij} \le y_j \quad \forall i,j \quad (9.3)$$

$$\sum_j a_j y_j + \sum_i \sum_j b_{ij} x_{ij} + d_1^- - d_1^+ = F \quad (9.4)$$

$$\sum_j p_{ij} x_{ij} + d_{2i}^- - d_{2i}^+ = D_i \quad \forall i \quad (9.5)$$

$$\sum_j (1 - \theta_j) y_j + d_3^- - d_3^+ = \theta^0 \quad (9.6)$$

$$x_{ij} = 0 \, \forall i \in \Delta_k \quad \text{and} \quad \alpha_{jk} = 0 \quad (9.7)$$

$$\sum_j x_{ij} = 1 \quad \forall i \quad (9.8)$$

$$x_{ij} \in \{0,1\},\; y_j \in \{0,1\} \quad \forall i,j,t \quad (9.9)$$

$$d_1^-, d_1^+ \ge 0, \quad d_{2i}^-, d_{2i}^+ \in \{0,1\}, \quad d_3^-, d_3^+ \ge 0 \quad (9.10)$$

The objective function in Eq. 9.1 represents the deviational variables to be optimized associated with the goal constraints given in Eqs. 9.4–9.6. Constraints in Eq. 9.2 represent the 'volume capacity' constraints which ensure that the total volume of containers assigned to the particular shipper does not exceed its maximum capacity. Constraints in Eq. 9.3 forces $y_j = 1$ when a shipper is selected. Constraint in Eq. 9.4 represents the 'fund availability' goal constraint which ensures that the total expenditure should not exceed the available fund. The first term in Eq. 9.4 represents the fixed costs associated with the selected shippers and the second term represents the variable costs for shipping the containers through a particular shipper. The 'due date delivery' goal constraints in Eq. 9.5 ensure that every container should be delivered to the customer on or before the stipulated delivery date. Constraint in Eq. 9.6 ensures that number of shippers selected should not exceed the maximum allowable non-compliant shippers. Constraints in Eq. 9.7 ensure that a container is not shipped through a shipper that cannot handle the type of goods in the container. Constraints in Eq. 9.8 ensure that each container is shipped through exactly one shipper.

9.1.2 Multi Period Model

We define the following (binary) decision variables:

$$\hat{x}_{ijt} = \begin{cases} 1 & \text{if container } i \text{ is shipped through shipper } j \text{ in period } t \\ 0 & \text{otherwise} \end{cases}$$

$$y_j = \begin{cases} 1 & \text{if shipper j is chosen} \\ 0 & \text{otherwise} \end{cases}$$

$$\hat{y}_{jt} = \begin{cases} 1 & \text{if shipper j is chosen in period } t \\ 0 & \text{otherwise} \end{cases}$$

The integer linear programming is

$$Min \sum_t w_1 \hat{d}_{1t}^+ + \sum_i w_{2i} d_{2i}^+ + w_{31} d_3^+ - w_{32} d_3^- \tag{9.11}$$

$$\sum_i e_i \hat{x}_{ijt} \leq \hat{c}_{jt} \quad \forall j, t \tag{9.12}$$

$$\hat{x}_{ijt} \leq \hat{y}_{jt} \quad \forall i, j, t \tag{9.13}$$

$$\hat{y}_{jt} \leq y_j \quad \forall j, t \tag{9.14}$$

$$\sum_j a_j \hat{y}_{jt} + \sum_i \sum_j b_{ij} \hat{x}_{ijt} + \hat{d}_{1t}^- - \hat{d}_{1t}^+ = \hat{F}_t \quad \forall t \tag{9.15}$$

$$\sum_t \sum_j (p_{ij} + t) \hat{x}_{ijt} + d_{2i}^- - d_{2i}^+ = D_i \quad \forall i \tag{9.16}$$

$$\sum_j (1 - \theta_j) y_j + d_3^- - d_3^+ = \theta^0 \tag{9.17}$$

$$\hat{x}_{ijt} = 0 \quad \forall\, t, \quad i \in \Delta_k \quad \text{and} \quad \alpha_{jk} = 0 \tag{9.18}$$

$$\sum_j \sum_t \hat{x}_{ijt} = 1 \quad \forall\, i \tag{9.19}$$

$$\hat{x}_{ijt} \in \{0,1\}, \quad y_j \in \{0,1\}, \quad \hat{y}_{jt} \in \{0,1\} \quad \forall\, i,j,t \tag{9.20}$$

$$d_{1t}^{-}, d_{1t}^{+} \geq 0, \quad d_{2i}^{-}, d_{2i}^{+} \in \{0,1\}, \quad d_{3}^{-}, d_{3}^{+} \geq 0 \tag{9.21}$$

Equation 9.11 represents the deviational variables to be optimized. Constraint in Eq. 9.12 represents the volume capacity constraints. Constraint in Eq. 9.14 forces $y_j = 1$ whenever shipper j is selected. Constraint in Eq. 9.13 forces $\hat{y}_{jt} = 1$ when shipper j is selected in period t. Constraint in Eq. 9.15 represents the 'fund availability' goal constraint. The 'due date delivery' goal constraints in Eq. 9.12 ensure that every container should be delivered to the customer on or before the stipulated delivery date. Constraint in Eq. 9.17 ensures that number of shippers selected does not exceed the maximum allowable non-compliant shippers. Constraint in Eq. 9.14 ensures that a container is not shipped through a shipper that cannot handle the type of goods in the container. Constraint in Eq. 9.19 ensures that every container is shipped through exactly one shipper on a particular period.

9.2 Numerical Experiments and Results

The CI procedure is now applied to solve the Selection of Cross-border Shippers (SCBS) problem discussed in Sect. 9.1. The procedure is coded in MATLAB 7.7.0 (R2008B). In addition, the simulations are run on a Windows platform using an Intel Core2 Quad CPU, 2.6 GHz processor speed and 4 GB memory capacity. In total, 8 distinct cases presented in Table 9.1 are solved for the single-period version and 18 cases presented in Table 9.2 are solved for the multi-period version of the problem. For every case, 10 instances are generated and every instance is solved 10 times using the CI method. The associated CI parameters such as the number of candidates S and the number of variations Y are chosen to be 3 and 10, respectively.

For all the considered problem cases, the number of different types of goods is set equal to $K = 5$. The size of the set of containers Δ_k for every type of good $k = 1, 2, \ldots, K$ chosen for every problem is listed in Tables 9.1 and 9.2 The value of α_{jk} randomly chosen to be either 0 or 1 such that each good $kk = 1, 2, \ldots, K$ is handled by at least one shipper $j = 1, 2, \ldots, J$. Each shipper $j, j = 1, 2, \ldots, J$, is randomly chosen to be either compliant $(\theta_j = 1)$ or noncompliant $(\theta_j = 0)$. The maximum allowable number of non-compliant shippers θ^0 are considered to be equal to the number of non-compliant shippers.

Furthermore, the fixed costs a_j for compliant and non-compliant shippers are uniformly generated from within the intervals $[100, 150]$ and $[150, 250]$,

Table 9.1 Results for test problems (single period)

J, I	Δ_k	N_v	N_c	N_{in}	IP	CI performance			% gap MRSLS versus IP	% gap MRSLS versus CI
					CPU time (s)	Avg sol % gap	CPU time (s)	SD (CPU time)		
5, 41	15, 10, 8, 5, 3	625	621	10	0.4867	2.5042	2.8276	1.0843	5.4395	4.5194
6, 47	17, 12, 9, 6, 3	869	873	10	0.4695	2.9914	2.2563	2.3568	3.5952	1.9336
8, 64	22, 18, 8, 10, 6	652	586	10	0.8511	4.1836	2.7992	0.7857	5.2474	2.8775
4, 132	40, 32, 25, 15, 20	800	666	10	0.5818	5.7075	7.3635	3.9597	10.1657	5.0322
8, 91	40, 35, 25, 15, 17	922	829	10	1.2820	4.6123	6.6243	2.4293	11.0711	6.0294
3, 143	50, 40, 20, 15, 18	1442	1297	10	5.6789	6.1791	10.6124	5.0364	10.9465	5.8111
8, 900	350, 250, 150, 100, 50	5349	904	10	42.9057	5.9488	30.4846	10.0735	11.1418	4.7671
8, 965	400, 250, 130, 100, 85	9662	8695	10	55.9057	5.9132	37.1572	7.2841	16.7324	8.2755

Table 9.2 Results for test problems (multi period)

T, J, I	Δ_k	N_v	N_c	N_{in}	IP	CI Performance			% gap MRSLS versus IP	% gap MRSLS versus CI
					CPU time (s)	Avg sol % gap	CPU time (s)	SD (CPU time)		
3, 5, 41	15, 10, 8, 5, 3	625	621	10	0.2293	4.2318	8.9319	2.5987	17.1880	13.7941
3, 6, 47	17, 12, 9, 6, 3	869	873	10	0.1778	2.6832	8.0473	3.5242	85.6929	82.4845
4, 3, 64	22, 18, 8, 10, 6	840	840	10	0.1107	2.1852	6.7515	2.5531	7.7782	5.5640
2, 4, 132	40, 32, 25, 15, 20	1181	1178	10	0.2932	5.0093	20.3730	6.3161	12.6099	8.9235
3, 3, 91	40, 35, 25, 15, 17	825	822	10	0.1560	2.5276	15.3088	2.9411	19.9835	19.1615
8, 3, 143	50, 40, 20, 15, 18	1032	1030	10	0.2230	2.1348	17.0945	3.4270	44.8077	42.6552
3, 4, 900	350, 250, 150, 100, 50	11104	11104	10	0.8642	5.6406	36.2087	7.8925	19.3918	14.4520
4, 8, 965	400, 250, 130, 100, 85	9662	8695	10	5.2541	3.8738	74.6240	14.1965	11.1537	7.2439
4, 25, 1000	300, 250, 200, 150, 100	27029	26026	10	40.232	6.7197	79.1075	18.0798	29.4734	23.8698
8, 3, 2871	900, 700, 600, 500, 171	66730	66734	10	9.9120	4.7621	73.0579	5.7850	53.4205	47.9887
8, 3, 3876	1400, 1000, 800, 500, 176	87946	87952	10	18.044	11.4148	108.0878	35.0878	52.4043	37.0317
5, 37, 1954	800, 500, 300, 200, 154	316461	316556	10	210.10	5.9746	113.6585	40.0708	102.0578	91.7036
9, 47, 1521	600, 400, 300, 121, 100	542795	543019	10	376.31	10.1997	93.0901	22.9238	55.1360	48.6519

(continued)

Table 9.2 (continued)

T, J, I	Δ_k	N_v	N_c	N_{in}	IP	CI Performance			% gap MRSLS versus IP	% gap MRSLS versus CI
					CPU time (s)	Avg sol % gap	CPU time (s)	SD (CPU time)		
8, 15, 6576	3000, 2000, 800, 500, 276	883635	883705	10	528.07	11.5074	141.5355	10.6661	174.4285	139.5145
8, 15, 5286	2000, 1000, 986, 800, 500	567372	567434	10	352.62	9.8089	130.3254	11.6471	129.1648	103.3162
8, 13, 5479	300, 250, 200, 150, 100	493533	493586	10	399.50	10.8206	133.1693	23.9897	134.3845	112.7871
8, 8, 4954	1400, 1000, 800, 500, 176	276901	276923	10	128.39	11.5375	71.7315	16.1202	153.8094	128.5710
8, 26, 3249	900, 800, 700, 600, 249	581894	582007	10	410.17	10.2553	133.7576	10.8573	157.9798	135.3840

respectively. The variable costs b_{ij} of shipping container i through shipper j for compliant and non-compliant shippers are uniformly generated from within the interval $[20, 50]$ and $[50, 80]$, respectively. Similarly, The funds available F and $\hat{F}_t$ are uniformly generated from within the interval $\left[max(b_{ij}) + \frac{I}{4},\ max(b_{ij}) + \frac{I}{2}\right], i = 1, 2, \ldots, I$ and $j = 1, 2, \ldots, J$. The maximum capacities c_j and $\hat{c}_{jt}$ and the volumes e_i, $i = 1, 2, \ldots, I$, are uniformly generated from within the interval $[200, 900]$ and $[10, 25]$, respectively. In addition, the expected processing times p_{ij} of container i through shipper j for both compliant and non-compliant shippers are uniformly selected from within the interval $[T/2, T]$ and $[1, T]$, respectively. Finally, the due dates $D_i, i = 1, 2, \ldots, I,$ are randomly generated from within $\left[min(p_{ij}), (max(p_{ij}) + 1)\right], i = 1, 2, \ldots, I$ and $j = 1, 2, \ldots, J$. Note that all the goals are considered equally important and are assigned weights equal to 1.

The average CI solution for every case is compared with the associated Integer Programming (IP) solution obtained by using CPLEX. The IP could solve the single period problem up to number of shippers $j = 8$ and number of containers $I = 965$, i.e. with 9662 variables and 8695 constraints. The performance comparison of the IP and CI solution is presented in Tables 9.1 and 9.2 along with the graphical illustration in Figs. 9.1 and 9.2.

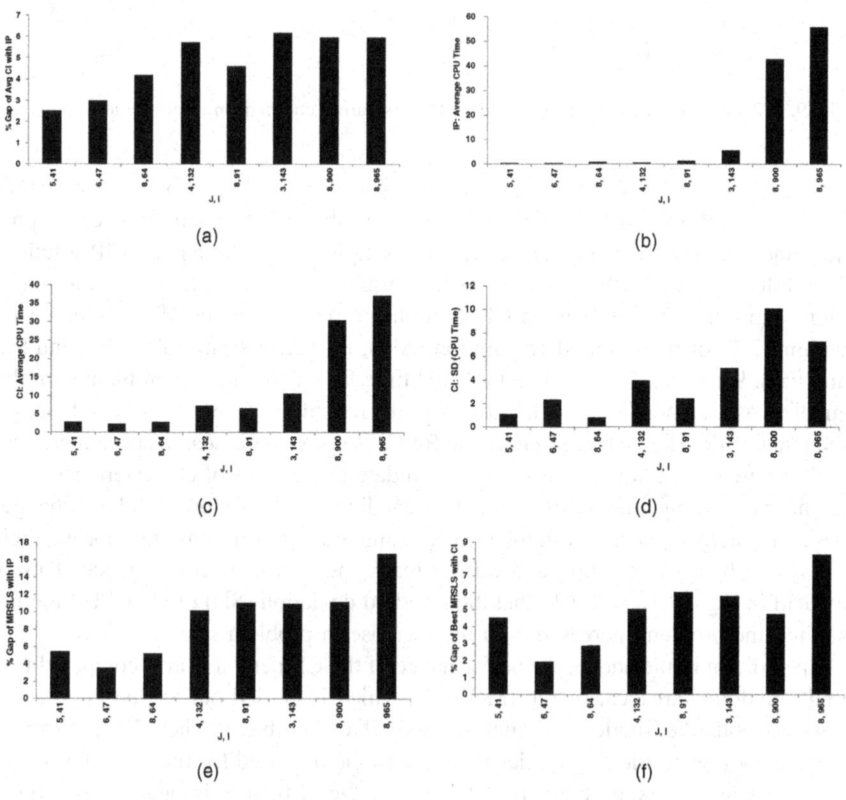

Fig. 9.1 Illustration of the CI, IP and MRSLS solution comparison (single period)

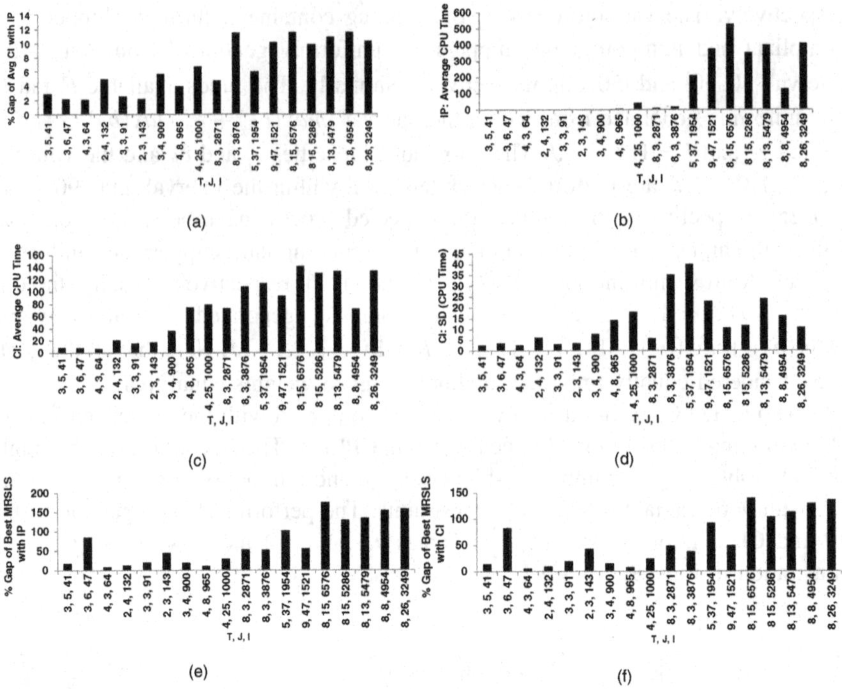

Fig. 9.2 Illustration of the CI and IP and MRSLS solution comparison (multi period)

It is evident from the results in Tables 9.1 and 9.2 and plot presented in Figs. 9.1a and 9.2a that for the smaller sized problems, the CI method could produce the solution comparatively closer, i.e. within 6 % of the reported IP solution. The difference gradually increased as the problem size grew; however, the maximum gap between the average CI solution and corresponding IP is noted to be within 12 % of the reported IP solution. Also, it is clear from Tables 9.1 and 9.2 and Figs. 9.1b, c and 9.2b, c that the CPU time for CI solving the problem with the smaller cases is more than the IP; however, the rate of increase is significantly lesser than that of IP. The increase in the time for CI is because the search space increased with problem size; however for every candidate the number of characteristics to be learnt in a learning attempt from the other candidate being followed did not change, which resulted in increased number of learning attempts and time to improve their individual behavior/solution and further reach the saturation/convergence. This is evident in Figs. 9.1d and 9.2d that the standard deviation (SD) of the CPU time for solving the problem increased with the increase in problem size.

In addition to the above, the performance of the CI method is also compared to a multi-random-start local search (MRSLS), which is carried out to find good solutions to both the single- and multi-period SCBS problem. The MRSLS implemented for this problem is similar in nature to the one used for finding solutions to the Sea Cargo Mix problem (see Chap. 8). Our MRSLS is again based on a pair-wise interchange argument to generate a neighboring solution from the one currently being assessed. More specifically, an initial solution is first constructed. This solution specifies an assignment of containers to shippers. To construct an alternative solution from the existing one, two shippers are randomly selected. Then, for each shipper, a subset of cargoes that are currently assigned to this shipper are randomly chosen. In the new solution, the selected cargoes are interchanged (or swapped) among the two designated shippers. This process is continued in every successive learning attempt until a stopping criteria is met.

For each of the case problems considered (Tables 9.1 and 9.2), the MRSLS for the single- and multi-period SCBS problems is run 50 times with different initializations. Also, for a meaningful comparison, every MRSLS case is initialized to start in the neighborhood of the CI's starting point and is run for exactly the same time equal to the corresponding average CPU time the CI method takes to solve that case. Similar to the (SCM) problem, the acceptance of the resulting solution in every learning attempt depends on the following feasibility-based rules [4]: (1) if the existing solution is infeasible and the resulting solution has improved constraint violation, then the solution is accepted, (2) If the existing solution is infeasible and the resulting solution is feasible, then the solution is accepted, (3) if the existing solution is feasible and the resulting solution is also feasible and the objective function has improved objective, then the solution is accepted. If any of these conditions are not satisfied then the existing solution is retained and the resulting solution is discarded.

It is important to mention here again that for many of the MRSLS runs carried out to completion for the single- and multi-period case problems, only few solutions are feasible and most of them do not satisfy the feasibility conditions. This is because for every MRSLS run a solution is randomly initialized which could be infeasible and MRSLS further might not have been able to discover a feasible solution. Therefore, only the best of the feasible solutions obtained using MRSLS are considered for comparison with the CI. From Tables 9.1 and 9.2 as well as from Figs. 9.1a, e, f and 9.2a, e, f it can be seen that the rate of increase of the percentage gap between the solution obtained using MRSLS and that obtained using CPLEX is significantly more when compared to the rate of percentage gap increase between CPLEX and CI. In addition, the percentage gap between the solution obtained using MRSLS and CPLEX per case is also considerably larger than the one achieved for CPLEX versus CI. In other words, CI has performed significantly better than MRSLS in finding good solutions to the SCBS problem.

9.3 Conclusions

The emerging optimization technique of cohort intelligence (CI) is successfully applied to solve a cross-border shippers' problem. The results indicate that the accuracy of solutions to these problems obtained using CI is fairly robust and the computational time is quite reasonable. Furthermore, the usefulness of CI in satisfactorily solving goal programming problems is also demonstrated.

The guiding principles of CI as an optimization procedure are grounded in artificial intelligence (AI) concepts. CI models the self-supervising behavior of a group of people seeking approximately the same goal. The self-supervising nature and rational behavior of the candidates among the cohort is illustrated along with the learning process that takes place among the candidates in order to further improve their individual characteristics/qualities. Furthermore, the inherent ability of the CI algorithm in handling complicated constraints lends to its applicability in solving real world complex problems. In addition, it is evident from the results that the variability as measured by standard deviation (SD) in the quality of solutions obtained using CI is commendable and remains almost stable as the problem size increases. This is because, even though the search space increases as the problem size increases, the number of characteristics in a learning attempt that need to be learnt by a candidate who is following the behavior of another candidate do not change. This results in an increase in the number of learning attempts in order to improve candidates' individual solutions and to finally reach the cohort's global solution.

Some limitations of the CI method should also be identified. The rate of convergence and the quality of the solution is dependent on the parameters such as the number of candidates and the number of variations. These parameters are derived empirically over numerous experiments and their calibration require some preliminary trials. It should also be observed that the number of characteristics attempted to adopt/learn is an important parameter when dealing with combinatorial optimization problems. As fewer characteristics are considered during the learning stage, this may delay the method's convergence rate significantly. The procedure may get stuck in the neighborhood of a local minimum, which may result into premature convergence. How to fine-tune the CI parameters and what to decide on the number of characteristics that needs to be learned by a candidate in every learning attempt can be done in an evolutionary and adaptive way as discussed in [5]. This may also help in increasing the accuracy of the solution as well as reducing the SD and overall performance of the algorithm. In addition, it should also be observed that the initial guess of the candidate solutions can affect the computational time of the algorithm. More specifically, if the initial candidate solutions are closer to the feasible region the chances of achieving saturation/convergence and reaching the optimal solution faster are high.

The paper also describes the application of a multi-random-start local search (MRSLS) that can be used to solve these three problems. The MRSLS implemented here is based on the interchange argument, a valuable technique often used in sequencing, whereby the elements of two adjacent solutions are randomly interchanged in the process of searching for better solutions. Our findings are that the performance of the CI is clearly superior to that of the MRSLS for many of the problem instances that have been solved.

As mentioned before, in agreement with the no-free-lunch theorem [6], any algorithm may not be directly applicable to solve all the problem types unless it can be enhanced by incorporating some useful techniques or heuristics. The CI method may also benefit from certain performance-enhancing techniques when it is applied to different classes of problems. A mechanism to solve multi-objective problems is currently being developed, which can prove helpful in transforming the model's constraints into objectives/criteria (see [4, 6] for new development in this area). This can help reduce the dependency on the quality of the candidates' initial guess.

References

1. Kulkarni, A.J., Baki, M.F., Chaouch, B.A.: Application of the cohort-intelligence optimization method to three selected combinatorial optimization problems. Eur J Oper Res. (2015)
2. Li, Z., Bookbinder, J.H., Elhedhi, S.: Optimal shipment for an airfreight forwarder: formulation and solution methods. Transp. Res. Part C **21**, 17–30 (2012)
3. Wong, W.H., Lawrence, C.L., Hui, Y.V.: Airfreight forwarder shipment planning: a mixed 0-1 model and managerial issues in the integration and consolidation of shipments. Eur. J. Oper. Res. **193**, 86–97 (2009)
4. Deb, K.: An efficient constraint handling method for genetic algorithms. Comput. Methods Appl. Mech. Eng. **186**(2–4), 311–338 (2000)
5. Eiben, A.E., Smit, S.K.: Evolutionary algorithm parameters and methods to tune them. In: Hamidi, Y., et al. (eds.) Autonomous Search, pp. 15–36. Springer, Berlin (2011)
6. Patankar, N.S., Kulkarni, A.J., Tai, K., Ghate, T.D., Parvate, A.R.: Multi-criteria probability collectives. Int. J. Bio-Inspired Comput. **6**(6), 369–383 (2014)

Chapter 10
Conclusions and Future Directions

This book provided detailed state-of-the-art developments on the emerging socio-inspired metaheuristic technique of Cohort Intelligence (CI). The motivation of the methodology is also discussed in detail. The methodology was successfully tested and validated by solving several unconstrained problems with different modalities and dimensions. The solution quality was quite promising and encouraging in terms of objective function, robustness, avoidance of local minima, computational time and function evaluations. The effect of each individual parameter such as sampling interval reduction factor, number of candidates and number of variations on the computational performance was also tested.

The book also validated the constraint handling ability of the CI methodology by solving a variety of well known test problems including three mechanical engineering design problems. The objective functions were of type polynomial, quadratic, cubic and nonlinear. A penalty function approach was incorporated for handling the constraints. In all the problem solutions, the implemented CI methodology produced sufficiently robust results with reasonable computational cost. This also justified the possible application of CI for solving a variety of real world problems.

The CI algorithm has been applied for solving several cases of five combinatorial NP-hard problems. The 0–1 Knapsack Problem (KP), with number of objects varying from 4 to 75 was the first one. In all the associated cases, the implemented CI methodology produced satisfactory results with reasonable computational cost. Furthermore, according to the solution comparison of CI with other contemporary methods it could be seen that the CI solution is comparable and for some problems even better than the other methods. In addition, in order to avoid saturation of cohort at suboptimal solution and further make the cohort saturate to the optimum solution, a generic approach such as accepting random behavior was incorporated.

Furthermore, the CI algorithm has been applied for solving combinatorial NP-hard Traveling Salesman Problem (TSP) with number of cities varying from 14 to 29. The application of the CI methodology for solving combinatorial NP-hard problem such as the TSP is successfully demonstrated. The CI incorporated with

A.J. Kulkarni et al., *Cohort Intelligence: A Socio-inspired Optimization Method*,
Intelligent Systems Reference Library 114, DOI 10.1007/978-3-319-44254-9_10

the roulette wheel approach, best behavior selection as well as random behavior selection approaches was successfully proposed. It is demonstrated that always following the best behavior/solution may make the cohort to saturate faster; however may make the cohort stuck into local minima. In addition, in order to jump out of possible local minima and further make the cohort saturate to global minimum, a generic approach such as accepting worst behaviors was incorporated. The encouraging results may help solve the real world problems with increasing complexity as the TSP can be further generalized to a wide variety of routing and scheduling problems [1]. In addition, CI approach could be modified to make it solve Multiple TSP (MTSP) and Vehicle Routing Problem (VRP). In this context, author see potential real world applications related to the distributed communication system such as, path planning of Unmanned Aerial vehicles (UAV) and addressing the ever growing traffic control problem using Vehicular ad hoc network (VANET).

In addition to above NP-hard problems, the CI was successfully applied to solve the new variant of the assignment problem, which has applications in healthcare and supply chain management. The results indicate that the accuracy of solutions to these problems obtained using CI is fairly robust and the computational time is quite reasonable. The results were compared with the multi-random-start local search (MRSLS) method. Moreover, several cases of the complex combinatorial problem such as the sea cargo mix problem were also successfully solved. The results were also compared with the MRSLS implemented. The findings are that the performance of the CI is clearly superior to that of Integer Programming (IP), specially developed heuristics referred to as HAM and MHA as well as the MRSLS for most of the problem instances that have been solved. Furthermore, CI was successfully applied to solve a large sized cross-border shippers' problem. The results indicate that the accuracy of solutions to these problems obtained using CI is fairly robust and the computational time is quite reasonable. Furthermore, the usefulness of CI in satisfactorily solving goal programming problems is also demonstrated. It is important to mention here that while solving the combinatorial problems an inbuilt probability based constraint handling approach was revealed and deployed to drive the solution towards the feasible region and further improve.

The book also in detail described the application of a MRSLS that can be used to solve the above three problems. The MRSLS implemented here is based on the interchange argument, a valuable technique often used in sequencing, whereby the elements of two adjacent solutions are randomly interchanged in the process of searching for better solutions. Our findings are that the performance of the CI is clearly superior to that of the MRSLS for many of the problem instances that have been solved.

As CI exhibited great potential to solve a variety of optimization problems including for data clustering. However, in the preliminary experiments solving unconstrained test problems, it was observed that as the problem size increased, CI may converge slowly and prematurely to local optima. With the purpose of assuaging these drawbacks modified CI (MCI) was proposed. It outperformed CI in terms of both quality of solutions and the convergence speed. In addition, a novel hybrid K-MCI algorithm for data clustering was also proposed. This new algorithm exploited the merits of the two algorithms simultaneously. This combination of K-means and MCI allowed our proposed algorithm to convergence more quickly and prevented it from falling to local optima. The proposed method can be considered as an efficient and reliable method to find the optimal solution for clustering problems. In this research, the number of clusters was assumed to be known a priori when solving the clustering problems. Therefore, we can further modify our algorithm to perform automatic clustering without any prior knowledge of number of clusters. We may combine MCI with other heuristic algorithms to solve clustering problems, which can be seen as another research direction. Finally, our proposed algorithm may be applied to solve other practically important problems such as image segmentation [2], dispatch of power system [3].

It is important to mention that the guiding principles of CI as an optimization procedure are grounded in Artificial Intelligence (AI) concepts. CI models the self-supervising behavior of a group of people seeking approximately the same goal. The self-supervising nature and rational behavior of the candidates among the cohort is illustrated along with the learning process that takes place among the candidates in order to further improve their individual characteristics/qualities. Furthermore, the inherent ability of the CI algorithm in handling complicated constraints lends to its applicability in solving real world complex problems. In addition, it is evident from the results that the variability as measured by standard deviation (SD) in the quality of solutions obtained using CI is commendable and remains almost stable as the problem size increases. This is because, even though the search space increases as the problem size increases, the number of characteristics in a learning attempt that need to be learnt by a candidate who is following the behavior of another candidate do not change. This results in an increase in the number of learning attempts in order to improve candidates' individual solutions and to finally reach the cohort's global solution.

Some limitations of the CI method should also be identified. The rate of convergence and the quality of the solution is dependent on the parameters such as the number of candidates and the number of variations and reduction factor. These parameters are derived empirically over numerous experiments and their calibration requires some preliminary trials. It should also be observed that the number of characteristics attempted to adopt/learn is an important parameter when dealing with combinatorial optimization problems. As fewer characteristics are considered during the learning stage, this may delay the method's convergence rate significantly. The procedure may get stuck in the neighborhood of a local minimum, which may result into premature convergence. How to fine-tune the CI parameters and what to decide on the number of characteristics that needs to be learned by a candidate in every learning attempt can be done in an evolutionary and adaptive way as discussed in [4]. This may also help in increasing the accuracy of the solution as well as reducing the SD and overall performance of the algorithm. In addition, it should also be observed that the initial guess of the candidate solutions can affect the computational time of the algorithm. More specifically, if the initial candidate solutions are closer to the feasible region the chances of achieving saturation/convergence and reaching the optimal solution faster are high.

References

1. Somhom, S., Modares, A., Enkawa, T.: Competition-based neural network for the multiple travelling salesman problem with Minmax objective. Comput. Oper. Res. **26**(4), 395–407 (1999)
2. Bhandari, A.K., Singh, V.K., Kumar, A., Singh, G.K.: Cuckoo search algorithm and wind driven optimization based study of satellite image segmentation for multilevel thresholding using kapurs entropy. Expert Syst. Appl. **41**, 3538–3560 (2014)
3. Zhisheng, Z.: Quantum-behaved particle swarm optimization algorithm for economic load dispatch of power system. Expert Syst. Appl. **37**, 1800–1803 (2010)
4. Eiben, A.E., Smit, S.K.: Evolutionary algorithm parameters and methods to tune them. Hamidi, Y., et al. (Eds.) Autonomous Search, pp. 15–36. Springer, Berlin (2011)

Printed in the United States
By Bookmasters